Ultraviolet Astronomy and the Quest for the Origin of Life

Ultraviolet Astronomy and the Quest for the Origin of Life

Edited by

Ana I. Gómez de Castro

*Professor in Astronomy and Astrophysics
and member of the Mathematics Faculty,
Universidad Complutense de Madrid (UCM), Spain*

ELSEVIER

Elsevier
Radarweg 29, PO Box 211, 1000 AE Amsterdam, Netherlands
The Boulevard, Langford Lane, Kidlington, Oxford OX5 1GB, United Kingdom
50 Hampshire Street, 5th Floor, Cambridge, MA 02139, United States

Notices
Knowledge and best practice in this field are constantly changing. As new research and experience broaden our understanding, changes in research methods, professional practices, or medical treatment may become necessary.

Practitioners and researchers must always rely on their own experience and knowledge in evaluating and using any information, methods, compounds, or experiments described herein. In using such information or methods they should be mindful of their own safety and the safety of others, including parties for whom they have a professional responsibility.

To the fullest extent of the law, neither the Publisher nor the authors, contributors, or editors, assume any liability for any injury and/or damage to persons or property as a matter of products liability, negligence or otherwise, or from any use or operation of any methods, products, instructions, or ideas contained in the material herein.

Library of Congress Cataloging-in-Publication Data
A catalog record for this book is available from the Library of Congress

British Library Cataloguing-in-Publication Data
A catalogue record for this book is available from the British Library

ISBN: 978-0-12-819170-5

For information on all Elsevier publications visit our website at
https://www.elsevier.com/books-and-journals

Publisher: Candice Janco
Acquisitions Editor: Peter J. Llewellyn
Editorial Project Manager: Lena Sparks
Production Project Manager: Kiruthika Govindaraju
Cover Designer: Alan Studholme

Typeset by TNQ Technologies

Contents

CHAPTER 5 **UV facilities for the investigation of the origin of life** 115

Ana I. Gómez de Castro, Martin A. Barstow, Noah Brosch,
Patrick Coté, Kevin France, Sara Heap, John Hutchings,
S. Koriski, Jayant Murthy, Coralie Neiner, Aki Roberge,
Julia Román-Duval, Jason Rowe, Mikhail Sachkov,
Evgenya Schkolnik and Boris Shustov

Contributors

Martin A. Barstow
School of Physics and Astronomy, University of Leicester, Leicester, United Kingdom

Noah Brosch
Tel Aviv University, Tel Aviv, Israel

Ada Canet
Joint Center for Ultraviolet Astronomy (JCUVA), Universidad Complutense de Madrid, Madrid, Spain; U.D. Astronomia y Geodesia, Facultad de CC Matemáticas, Universidad Complutense de Madrid, Madrid, Spain

Patrick Coté
National Research Council, Herzberg Astronomy and Astrophysics Research Centre, Victoria, BC, Canada

Raissa Estrela
Jet Propulsion Laboratory, California Institute of Technology, Pasadena, CA, United States; Center for Radioastronomy and Astrophysics Mackenzie, Mackenzie Presbyterian University, São Paulo, Brazil

Kevin France
Laboratory for Atmospheric and Space Physics, University of Colorado, Boulder, CO, United States

Ana I. Gómez de Castro
Joint Center for Ultraviolet Astronomy (JCUVA), Universidad Complutense de Madrid, Madrid, Spain; U.D. Astronomia y Geodesia, Facultad de CC Matemáticas, Universidad Complutense de Madrid, Madrid, Spain

Sara Heap
University of Maryland, College Park, MD, United States

John Hutchings
National Research Council, Herzberg Astronomy and Astrophysics Research Centre, Victoria, BC, Canada

S. Koriski
Tel Aviv University, Tel Aviv, Israel

Jeffrey L. Linsky
JILA, University of Colorado and NIST, Boulder, CO, United States

Jayant Murthy
Indian Institute of Astrophysics, Bangalore, Karnataka, India

Coralie Neiner
Observatoire Paris-Meudom, Meudom, France

Aki Roberge
Goddard Space Flight Center – NASA, Greenbelt, MD, United States

Julia Román-Duval
Space Telescope Science Institute, Baltimore, MD, United States

Jason Rowe
Bishop's University, Sherbrooke, QC, Canada

Mikhail Sachkov
Joint Center for Ultraviolet Astronomy (JCUVA), Universidad Complutense de Madrid, Madrid, Spain; Institute of Astronomy of the Russian Academy of Sciences, Moscow, Russia

Evgenya Schkolnik
School of Earth and Space Exploration, Arizona State University, Phoenix, AZ, United States

Boris Shustov
Joint Center for Ultraviolet Astronomy (JCUVA), Universidad Complutense de Madrid, Madrid, Spain; Institute of Astronomy of the Russian Academy of Sciences, Moscow, Russia

Adriana Valio
Center for Radioastronomy and Astrophysics Mackenzie, Mackenzie Presbyterian University, São Paulo, Brazil

N. Chandra Wickramasinghe
Buckingham Centre for Astrobiology, University of Buckingham, Buckingham, United Kingdom

Dayal T. Wickramasinghe
Mathematical Sciences Institute, The Australian National University, Canberra, ACT, Australia

Foreword

The concept for this book was born at the Scientific Assembly of the Committee on Space Research (COSPAR) in Pasadena, in July 2018. Big conferences such as the COSPAR Assembly attract thousands of participants, in this case scientists and engineers from the space sector. Elsevier was also there, and editor Marisa LaFleur contacted me to edit a book on "Ultraviolet astronomy and the quest for the origin of life," the title of the event I was organizing. This book is the result of that challenge.

Understanding how life emerged on Earth is a major aim of science. Laboratory and astronomical investigations carried during the last 30 years have proven that the building blocks of life can be produced in space, on the icy coatings of dust grains, laying the grounds to the *soft panspermia theory*.

The *panspermia theory* argues that life is originated in space, in spatial ices, and continuously distributed to the planets by comets and meteorites. The soft panspermia theory stays one step behind. According to this, instead of living forms, amino acids, sugars, and the molecules required to form RNA are produced in space. Experiments consisting in the irradiation of interstellar ice analogs with stellar-like UV radiation have shown that indeed, the building blocks of the RNA can be produced in space. Moreover, these laboratory results agree well with the measurements obtained from meteorites and the data gathered by the Rosetta mission from comet C67/ Churyumov-Gerasimenko. The implications of this *soft* theory are enormous since, as a result, life forms in the Cosmos would be compatible at the molecular level.

Astronomers are deeply involved in the investigation of the emergence of life in the Universe. They are working at different levels: from the investigation of the formation and remote detection of life-related molecules (e.g., amino acids, sugars, and urea) to studying planet formation and searching for exoplanets with signatures of life. Stars and planetary systems form from dense molecular clouds, which are opaque to the interstellar UV radiation field. Within the clouds, interstellar dust grains stick together into larger grains that grow thick ice mantles. Large organic molecules such as the polycyclic aromatic hydrocarbons (PAHs) get depleted from the gas phase onto the grain surface. As gravitation proceeds and stars form, sources of UV radiation are generated within the clouds that affect the surface chemistry of the circumstellar dust and the very evolution of the protostellar disks into planetary systems.

UV radiation also plays a major role on the stability of planetary atmospheres and affects the survival of early living forms. UV irradiation can denature DNA by producing strand breakage; it also may generate mutagenic photoproducts that inhibit the transcription/replication of chromosomes. At the time life first appeared on Earth, 3.7 Gyr ago, there was no protecting ozone layer, and the first microorganisms had to develop strategies to shield from it. Today, we know that some iron-bearing molecules act as good sunscreen molecules and that some living forms can survive harsh space conditions. For instance, the LICHENS experiment has shown that some lichens from Sierra de Gredos, in Spain, can afford 2 weeks of full exposure to space (including solar UV radiation) and recover nearly the same photosynthetic activity upon return to Earth, with no significant structural changes.

Astrophysical processes leave their imprint in the electromagnetic spectrum. The resonance transitions of the most abundant species in the universe and the electronic transitions of the most abundant molecules are in the UV range. For this reason, UV instrumentation is of prime importance for astrobiological research. The new generation of cubesats offers ample opportunities to test many new ideas and instruments. Their moderate price, versatility and off-the-shelf policy is paving the path to this revolution. The standardization of the cubesat components and the development of synergetic networks is fundamental for its success.

The chapters of this book have been chosen to provide an ample view of the ongoing research in the field; in a sense, this book can be taken as an up-to-date introduction to the field.

Chapter 1 is devoted to describe the foundations of the panspermia theory, as it was originally put forward in 1974 by Fred Hoyle and Chandra Wickramasinghe. Results from space missions, such as Stardust and Rosetta, are described under this light.

Chapter 2 reviews the current status of the investigation of the origin of life at UV wavelengths. The role that UV radiation has in the formation of planetary systems, planets, and eventually, life is also addressed.

Chapter 3 focusses on the impact of the stellar flaring activity in the survival of microorganisms on the planetary surfaces and the oceans.

Chapter 4 provides a detailed view of the environment where the Solar System resides in the Galaxy and where Earth-like exoplanets are being searched for.

Chapter 5 is a collective work that summarizes the status and characteristics of all the space projects, either operational, under development, or being proposed, that carry instrumentation for the investigation of origin of life at UV wavelengths.

Finally, I would like to thank Marisa LaFleur and Lena Spark, the Elsevier editors, for their support during these years. They inspired this book and accompanied us during this venture, showing an astronomically sized patience. It has been a great pleasure working with you.

Ana Inés Gómez de Castro
Universidad Complutense de Madrid
Madrid, January 2021

The growing case for life as a cosmic phenomenon

N. Chandra Wickramasinghe[1], Dayal T. Wickramasinghe[2]

[1]*Buckingham Centre for Astrobiology, University of Buckingham, Buckingham, United Kingdom;*
[2]*Mathematical Sciences Institute, The Australian National University, Canberra, ACT, Australia*

1. Introduction

Fred Hoyle and one of us (NCW) published the original proposal of graphite particles as interstellar grains as far back as 1962 (Hoyle and Wickramasinghe, 1962), and also a model for interstellar extinction involving mixtures of graphite and silicate grains in 1965 and 1969 (Wickramasinghe and Guillaume, 1965; Lynds and Wickramasinghe, 1968; Hoyle and Wickramasinghe, 1969). Although these ideas initially faced resistance, they are now widely accepted and often used uncritically as the only viable model of interstellar grains (Draine, 2003). We show here that the use of a graphite particle model to account for interstellar extinction, in particular the broad absorption band centered on 2175 Å, involves assumptions that are probably unrealistic. We argue that interstellar extinction observations, and other more recent astronomical data relating to comets, all support a biological model in preference to a wide class of abiotic or inorganic models. The main objection to a biological hypothesis appears to be cultural rather than scientific.

2. Interstellar extinction

From the 1970s onwards data on the interstellar extinction curve have been extended both with respect to the wavelength covered as well as the distances of sources examined. The wavelength dependence of extinction for wavelengths long ward of 3000 Å is known to be more or less invariant compared to the extinction for shorter ultraviolet wavelengths (Hoyle and Wickramasinghe, 1991). The main variation in the ultraviolet is with regard to slope as well as the extent of a broad symmetric hump in extinction centered at 2175 Å. Extinction curves for Large Magellanic Cloud (LMC) and Small Magellanic Cloud (SMC) have also been known since 1980, the former found to be identical to the galactic curve and the latter different to the extent that it possessed a weak or nonexistent 2175 Å bump. Extinction curves have more recently been extended to distant galaxies using a variety of techniques including far infrared observations of red-shifted galaxies including gamma ray burst (GRBs) galaxies) (Motta et al., 2002; Scoville et al., 2015; Noll et al., 2009; Catzetti et al., 2000).

Ultraviolet Astronomy and the Quest for the Origin of Life. https://doi.org/10.1016/B978-0-12-819170-5.00001-4

Fig. 1.1A shows the extinction curve for the galaxy SBS0909 + 532 at redshift $z = 0.83$ obtained by Motta et al. (2002) which is very similar to the galactic extinction curve. Fig. 1.1B shows a compilation by Scoville et al. (Scoville et al., 2015) of extinction curves for galaxies up to $z = 6$, while Fig. 1.1C shows the extinction data for galaxies at $z = 2$ compiled by Noll et al. (Noll et al., 2009) from the GMASS ultradeep survey. We note here that Catzetti et al. (Catzetti et al., 2000) have found only a weak or nonexistent 2175 Å feature in a group of local starburst galaxies, very similar to the situation for SMC.

It is generally considered that the most plausible explanation for the 2175 Å feature of interstellar dust involves a graphite particle model (Hoyle and Wickramasinghe, 1962; Wickramasinghe and Guillaume, 1965). Wickramasinghe and Guillaume (Wickramasinghe and Guillaume, 1965) used laboratory optical constants of graphite by obtained by Taft and Phillip (Taft and Phillipp, 1965) together with Mie formulae (van de Hulst, 1957) to compute the extinction cross-sections of graphite spheres of various radii. All the subsequent calculations and claims for a graphite grain model have followed the same trend.

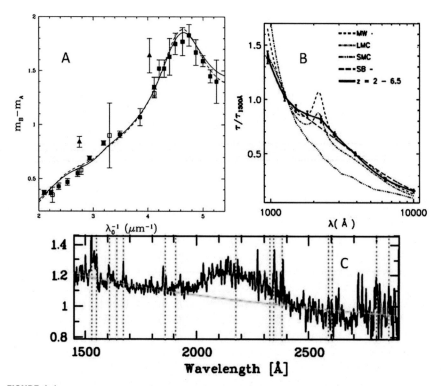

FIGURE 1.1

Extinction curves for galaxies of various redshift. Milky Way (MW) (Seaton, 1979), LMC (Fitzpatrick, 1986), SMC (Prevot et al., 1984), and local starburst galaxies (Catzetti et al., 2000). Compilation is due to Scoville (Scoville et al., 2015).

3. Failure of graphite grain model

The property of graphite that is needed to explain the 2175 Å humps in Fig. 1.1 arises from π to π^* electronic transitions that show up as a resonance in the complex refractive index $m(\lambda)$. Graphite consists of stacks of covalently bonded C_6 hexagons leading to a highly anisotropic crystal structure. Electrical conductivity and optical constants for electric vectors parallel and normal to the crystal planes differ markedly. The electrical conductivity parallel to the crystal planes are about 100 times greater than in the perpendicular direction; the bulk of the absorption may accordingly be assumed to come from electric vector $\overrightarrow{E}$ parallel to crystal planes.

The optical constants used in all the existing astronomical calculations for graphite grains assume that data for $\overrightarrow{E}$ parallel to the basal planes hold isotropically. Such a spherical isotropic model was assumed to be analogous to one where anisotropic flakes are randomly oriented but no experiment justifying this premise has so far been carried out. We now examine the validity of this hypothesis using a straightforward theoretical calculation.

A graphite flake comprised of a stack of planar hexagonal layers may be well represented by a thin circular disk of radius a, and thickness $t \ll a$. The complex dielectric constant for light with electric vector in the plane of the disk is obtained from measurements of $n_\parallel$ and $k_\parallel$ (Taft and Phillipp, 1965) as

$$\varepsilon_\parallel = m_\parallel^2 = \left(n_\parallel - ik_\parallel \right)^2 \qquad (1.1)$$

where m is the complex refractive index, n is the real part of the refractive index, and k is the absorptive index. The complex dielectric constant for electric vector transverse to the planes may be approximated by

$$\varepsilon_\perp = m_\perp^2 = K - 2i\sigma_\perp \lambda/c \qquad (1.2)$$

where $K = 4$ and $\sigma_\perp = \sigma_\parallel/100$ and $\sigma_\parallel$ is the conductivity for light with electric vector along the planes. Because $\sigma_\parallel/c = n_\parallel k_\parallel$ Eq. (1.1) yields

$$\varepsilon_\perp = 4 - 2in_\parallel(\lambda)\kappa_\parallel(\lambda)/100 \qquad (1.3)$$

In the Rayleigh limit where $\lambda \gg 2\pi a$, the extinction cross-section for light with electric vector parallel to the plane of the disk is

$$C_\parallel = -\frac{2\pi V}{\lambda} Im\left(\varepsilon_\parallel - 1 \right) \qquad (1.4)$$

and that for light with electric vector perpendicular to the plane is

$$C_\perp = -\frac{2\pi V}{\lambda} Im\left(\frac{\varepsilon_\perp - 1}{\varepsilon_\perp} \right) \qquad (1.5)$$

where V is the particle volume (Wickramasinghe, 1973). For a randomly oriented ensemble of graphite flakes in the Rayleigh limit we thus have an average cross-section for extinction given by

$$C(\lambda) = \left(C_{\parallel} + 2C_{\perp} \right)/3 \qquad (1.6)$$

Normalized extinction efficiencies computed from Eq. (1.6) for a random distribution of Rayleigh scattering discs and from the Mie formula for spheres comprised of graphite are shown in Fig. 1.2.

For spherical particles we assume that the appropriate optical constants are those for $E_{\parallel}$ hexagonal planes. We note that the required wavelength position of the 2175 Å feature to account for the astronomical data demands a radius close to $0.02\ \mu$m, or a mean radius of a similar value, for a size distribution of spheres. Any significant departure from this preferred radius removes the desired correspondence to the astronomical data. Fig. 1.2 also shows that smaller spheres (Rayleigh approximation $2\pi a/\lambda \ll 1$) or discs or flakes in random orientation do not provide the required agreement. This problem has been known for some time, although it has been generally ignored by astronomers involved in modeling interstellar grains (Draine, 2003).

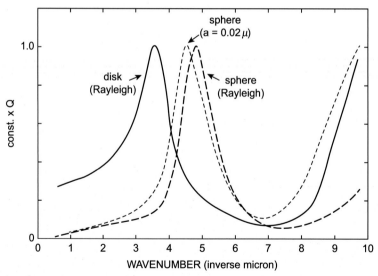

FIGURE 1.2

Solid curve is the normalized extinction computed for a random orientation of anisotropic graphite flakes calculated from Eqs. (1.1)–(1.6) with optical data of Taft and Phillipp (Taft and Phillipp, 1965); dashed curves are for spheres of the astronomically preferred radius $a = 0.02\ \mu$m, and also for the spheres in the Rayleigh limit.

4. Molecular hypothesis for the 2175 Å band

The case on which the canonical graphite model of interstellar dust is based, in our view, on very shaky grounds. It is for this reason that alternative molecular absorbers associated with invariant organic or biological grain ensembles have significant advantages. Polyaromatic hydrocarbons (PAHs) that are known to be present in the diffuse interstellar medium at first sight offer a better option to explain an ultraviolet absorption feature arising from π to π^* transitions in C_6 structures. The PAH molecules, however, vary in their ultraviolet absorption wavelengths depending on precise structure, size of cluster and varying ionization states. Unless an interstellar PAH ensemble can be defined uniquely and formed everywhere without *significant* variation the observed constancy of the 2175 Å absorption feature would be exceedingly difficult to understand (Hoyle and Wickramasinghe, 1991). Some attempts have been made in this direction but the ad hoc nature of proposed laboratory mixtures and their arbitrary ionizations states present a problem. Similar difficulties face the proposal that buckyonion grains involving C_{60} structures in the form of concentric shells of onion-like hyperfullerenes (Chhowalla et al., 2003).

For graphite spheres of optimal size the average mass density of grains required to produce the observed interstellar extinction excess of 1.5 mag kpc^{-1} over and above an underlying scattering background is $\sim 10^{-27}$ g cm^{-3} and the same result is found to hold for aromatic molecules (Hoyle and Wickramasinghe, 1977). This is not expected to be significantly different for the buckyonion case where the π-electrons are also involved. Thus, the mass fraction of interstellar carbon required in order to produce the 2175 Å absorption in whatever model one chooses, graphite spheres, aromatic molecules or buckyonions, can be shown to be of the general order of 10%. Accordingly we need to rely on an exceedingly efficient mode of production of buckyonion, or other alternative carriers.

A much better option to an arbitrarily constructed ensemble of PAHs or buckyonions would be to consider a highly reproducible biological system a bacterial or algal cell as the source of an interstellar aromatic ensemble for explaining the 2175 band and other diffuse interstellar bands (DIBs). In a recent study of red rain particles (algae of unknown identity in the 2001 red rain of Kerala). Kumar et al. (2019) have shown correspondences both in the 2175A extinction feature as well as in the DIBs.

5. Convergence to biology

One particularly important prediction of the biological model of interstellar dust is that the midinfrared spectrum of any infrared source seen through a few kilo parsecs of dust obscuration should reveal a predicted absorption spectrum of bacterial material. In 1981 this prediction was dramatically verified by David Allen and Dayal Wickramasinghe (Allen and Wickramasinghe, 1981; Wickramasinghe and Allen, 1980) for the galactic center source GC-IRS7 when it was shown that the spectrum was close to that predicted for partially desiccated bacteria (Hoyle et al., 1982a,b). This crucial fit is shown in the left-hand panel of Fig. 1.3. The result was interpreted as tenable evidence for life being a cosmic phenomenon (Rauf and Wickramasinghe, 2010; Hoyle et al., 1999; Hoyle et al., 1982a,b). Similar evidence accumulated in the subsequent 3 decades has served to strengthen this claim. Since 1980 the existence of complex organic molecules in interstellar clouds has also become indisputable.

The discovery of organic dust in comets came with the last perihelion passage of Comet Halley in 1986 (Wickramasinghe and Allen, 1986). The midinfrared spectrum (Fig. 1.3: right panel) of the comets dust coma following a major outburst on March 31, 1986 (Wickramasinghe and Allen, 1986) showed unambiguous evidence of aromatic-aliphatic linkages (C—H stretching modes) that were uncannily consistent with desiccated *E. coli* (Wickramasinghe et al., 1986). The solid curve is for nonirradiated bacteria; the dashed curve is for X-ray irradiated bacteria. More recent studies of other comets have yielded generally similar results.

Any alternative nonbiological explanation of the points in Fig. 1.3 would involve abiotically formed organic molecules possessing functional groups that fortuitously matched biology. This, in our view, is extremely improbable and poses a problem for abiotic explanations of the data.

Detections of interstellar organic molecules of ever-increasing complexity have continued in the last two decades following the deployment of newer and better instruments and telescopes. Infrared, microwave and radio observations are used to detect the presence of such molecules and the current list of positive detections is

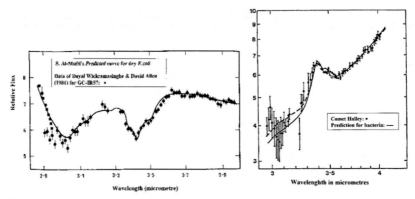

FIGURE 1.3

Left panel: Comparision of the normalized flux from GC-IRS7 obtained in 1980 (Allen and Wickramasinghe, 1981) with the laboratory spectrum of dessicated *Escherichia coli* (Hoyle et al., 1982a,b). Right panel: Emission by dust coma of Comet Halley observed by Dayal Wickramasinghe and David Allen on March 31, 1986 (points) (Wickramasinghe and Allen, 1986) compared with normalized fluxes for desiccated *E. coli* at an emission temperature of 320 K(Wickramasinghe et al., 1986).

likely to be constrained only by limitations of techniques. In addition to data on the ubiquitous 2175 Å ultraviolet absorption band discussed in Section 4, infrared, microwave and radio observations have shown the ubiquitous presence of organic molecules including PAHs. The infrared data are mainly in the form of absorptions at 3.3, 3.4, 6.2, 6.7, 8.6, 11.2, and 12.7 µm. They are present in galactic as well as extragalactic sources extending to redshifts of up to at least $z = 2$. The argument that these bands are independent of the 2175 Å dust feature is strengthened by observations of these bands in starburst galaxies, in which no 2175 Å band is seen (Catzetti et al., 2000). This is illustrated in the bottom frame of Fig. 1.4. The Spitzer telescope data (top frame) clearly show strong PAH features very similar in strengths and wavelengths to similar bands that occur in sources within the Milky Way. The data on the extinction curves of starburst galaxies, however, show no sign of any significant 2175 Å ultraviolet absorption. Although one might try to argue that PAH carriers of the 2175 Å bump are destroyed in shocks and intense radiation associated with starburst activity, this would not be consistent with the well-defined presence of 3.3, 3.4, 6.2, 6.7, 8.6, 11.2, and 12.7 µm features in the infrared. On the basis of our preferred biological hypothesis it could be argued that the absence of the 2175 Å feature indicates either no biological activity or a greatly diminished biology in starburst galaxies.

The totality of all the current available data points to small and intermediate-sized molecules resembling biochemicals. In addition we also have a set of diffuse interstellar absorption bands, typified by the 4430 Å band in the visual spectra of stars that have defied identification for over eight decades, which may also be due to biologically relevant molecules (Hoyle and Wickramasinghe, 1991).

The organic molecules thus far identified in interstellar clouds could be either degradation products of an all-pervasive cosmic microbiology or more conservatively they could be interpreted as steps toward the build-up or inception of biology.

6. More on comets

A largely organic composition of comets and the concept of cometary panspermia was proposed and developed by Hoyle and Wickramasinghe (Hoyle and Wickramasinghe, 1981). This concept (cometary panspermia) involves comets as the collectors, amplifiers, and redistributors of microbiota, the origin of which is posited to be a cosmological event.

As we have already noted, spectroscopic identification of interstellar dust and molecules in space started in the 1970s continues to come into ever sharper focus and their biochemical relevance, once contested, is now conceded. The trend remains, however, to assert without proof that we are witnessing the operation of prebiotic chemical evolution on a cosmic scale rather than biology. With no progress in the quest to make life from nonlife in the laboratory (Deamer, 2012), and considering the superastronomical odds against the spontaneous emergence of the simplest biological system (Hoyle and Wickramasinghe, 2000), astronomical data from both

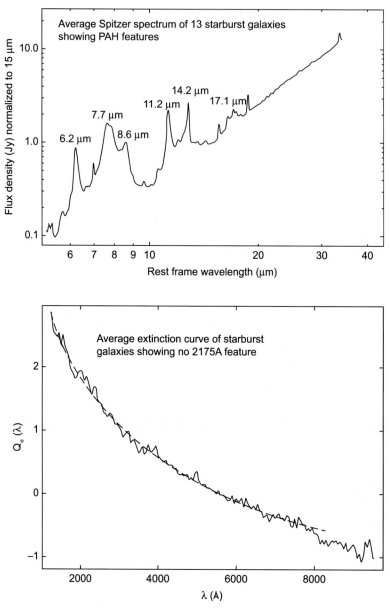

FIGURE 1.4

Top frame: average spectrum of starburst galaxies from (Brand et al., 2006) showing PAH bands; the bottom frame is the average extinction for starburst galaxies showing no evidence of a 2175 Å bump, and so removing the involvement of PAHs in the production of UV features in galaxies.

interstellar dust and comets point to biology operating on a galaxy-wide scale. If biological evolution and replication are regarded as the only reliable facts—life always generates new life—this must be so even on a cosmic scale.

The Rosetta Mission to comet 67P/C-G has continued to yield a wealth of data that satisfy consistency checks for biology. Fig. 1.5 shows the close consistency between the surface properties of the comet and the spectrum of a desiccated bacterial sample.

The presence of complex organic molecules including the building blocks of life in comets is now amply confirmed; so it is reasonable to hypothesize that there is fully fledged microbial life in comets. The reigning paradigm, however, firmly rejects this possibility. Life related chemicals and prebiotic molecules are at long last permitted but fully fledged life still appears taboo. Serious inconsistencies are beginning to arise. The Rosetta Missions Philae lander has recently provided novel information about the comet 67P/C-G (Capaccione et al., 2015). Jets of water and organics issuing from ruptures and vents in the frozen surface are consistent with biological activity occurring within subsurface liquid pools (Wallis and Wickramasinghe, 2015). The most recent report of O_2 along with evidence for the occurrence of water and organics provides further evidence of such ongoing biological activity (Bieler et al., 2015). Such a mixture of gases cannot be produced under

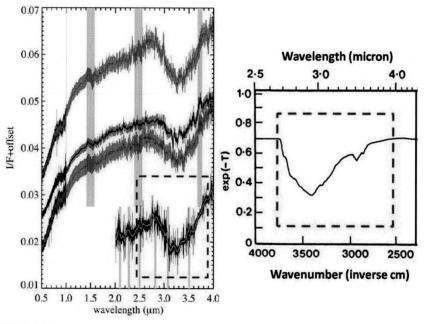

FIGURE 1.5

The surface reflectivity spectra of comet 67P/C-G (left panel) (Capaccione et al., 2015) compared with the transmittance curve measured for *E. coli* (right panel).

thermodynamic conditions, because organics are readily destroyed in an oxidizing environment. The freezing of an initial mixture of compounds, including O_2, not in thermochemical equilibrium, has been proposed, but there is no evidence to support such a claim. On the other hand the oxygen/water/organic outflow from the comet can be explained on the basis of subsurface microbiology. Photosynthetic microorganisms operating at the low light levels near the surface at perihelion could produce O_2 along with organics. Many species of fermenting bacteria can also produce ethanol from sugars, so the recent discovery that Comet Lovejoy emits ethyl alcohol amounting to 500 bottles of wine per second may well be an indication that such a microbial process is operating (Biver et al., 2015).

Next we turn to the recent report of the presence of the amino acid glycine and an abundance of phosphorus in the coma of comet 67P/CG (Altwegg et al., 2016). Fig. 1.6 shows the atomic mass spectrum of the coma gas examined by the Rosetta Orbiter Spectrometer for Ion and Neutral Analysis (ROSINA) mass spectrometer. Very high ratios of $P/C \approx 10^{-2}$ are difficult to reconcile on the basis of volatilization of condensed material of solar composition with $P/C \approx 10^{-3}$, particularly when we might expect inorganic phosphorus to be mostly fixed in refractory minerals. On the other hand the P/C ratio of biomaterial is close to that implied by the ROSINA data and would be explained if the material in the coma started off as biomaterial virions and bacteria (Fagerbakke et al., 1996).

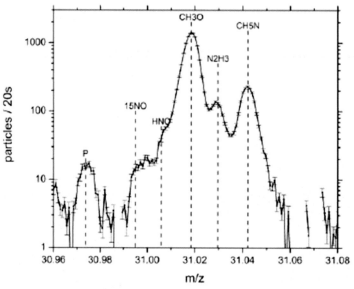

FIGURE 1.6

Phosphorus peak implying 1% phosphorus relative to carbon, consistent with biology and degradation of nucleic acids (Altwegg et al., 2016).

The idea that microbial life cannot exist in comets still dominates scientific orthodoxy but its rational basis is fast eroding (Wickramasinghe and Tokoro, 2014). Adhering to a paradigm that has long past its sell-by date will surely be to the detriment of science in the long run.

7. Icy moons

Apart from comets we have discussed elsewhere that microbial habitats could also be present in other icy bodies of the solar system. These include the icy moons of the giant planets such as Europa and Enceladus, with plausibly the bulk composition of their outer layers being water-ice. Such icy bodies, spanning a wide range in size between comets and icy moons of planets, can plausibly be considered the progenitors of larger icy planets (Hoyle and Wickramasinghe, 1981). All that such icy bodies contain not only water-ice and organic molecules which could serve as building blocks of life but also a residue of interstellar bacteria/diatoms/viruses that had retained viability over millions of years. Comets as well as larger icy bodies such as Enceladus at the time of their coalescence from smaller bodies, would have possessed an initial endowment of deep-frozen viable microbial life. After their formation into cometary or planetary bodies the energy released from the decay of the radioactive elements, such as ^{232}Th, ^{238}U, and ^{40}K, would have led to the development extensive liquid water domains beneath hundreds of meters of insulating crust for more than a Gyr (Hoyle and Wickramasinghe, 1978; Wallis, 1980; Yabushita, 1993). The heat flow outwards through several hundred meters of ice (with low thermal conductivity) would be so slow that liquid conditions and microbial habitability could persist for billions of years (Wallis and Wickramasinghe, 2015). With biological heat sources providing an additional source of heating, liquid water domains may well exist even in the present day.

Analyzing a slight wobble in the rotation of Enceladus about its axis, Thomas et al. (Thomas et al., 2015) inferred a present-day existence of a global liquid water ocean beneath a frozen icy crust. Similar arguments could be used to infer the presence of liquid water oceans in the Jovian moon Europa, and perhaps also in Pluto. In 2011 the Cassini spacecraft discovered jets of water, methane, dust, and salt crystals (NaCl) issuing at speeds exceeding $1000\ \mathrm{kms}^{-1}$ from vents in the south polar region of Enceladus (Spencer and Nimmo, 2013). These jets were described as curtain eruptions issuing through a network of linear fractures and fissures in the ice surface (Spitale et al., 2015). The curtains of issuing material travel outward at escape speed and are almost certainly driven by high gas pressures generated beneath the ice shell.

Recently (Posberg et al., 2018) evidence has been provided that such curtains of material actually contain particles of ice laden with high molecular weight complex organic molecules. Organic molecules with masses exceeding 200 AMU are within a small factor of biological molecules which typically have masses of perhaps a few thousand AMU. Within the scope of the new investigations of Enceladus that have yielded such amazing results, biology must surely come high up as a candidate for supplying such molecules. But this has yet to be conceded.

8. Geological evidence

Three decades ago the oldest evidence for microbial life in the geological record was thought to be in the form of cyanobacteria-like fossils dating back to 3.5 Gyr ago. From the time of the formation of a stable crust on the Earth 4.3 Gyr ago, during an episode of violent impacts by comets (the Hadean Epoch), there seemed then to be available a timespan of 800 Myr during which the canonical Haldane-Oparin primordial soup may have developed, leading ultimately to an origin of life on the Earth. However, very recent discoveries have shown that this time interval available for any such prebiotic process may have been effectively closed. Detrital zircons older than 4.1 Gyr, discovered in rocks belonging to a geological outcrop in the Jack Hills region of Western Australia, have been found to contain micron-sized graphite spheres which have an isotopic signature of biogenic carbon (Bell et al., 2015) (See Fig. 1.7). This signature results from the preference of biology to take up the lighter isotope ^{12}C rather than ^{13}C, an effect that can be seen in microbial fossil structures. Care should be taken, however, to exclude nonbiological effects that could mimic the same

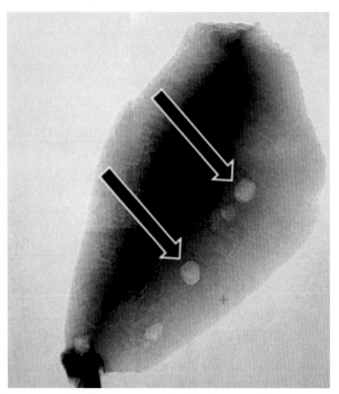

FIGURE 1.7

The ^{12}C enriched micron-sized graphite spheres showing possible evidence of degraded bacteria within zircon crystals (Bell et al., 2015).

phenomenon. For instance, the involvement of extrasolar grains with a high $^{12}C/^{13}C$ remains a possibility that needs to be discussed. Bell et al. (Bell et al., 2015) discuss nonbiological alternatives for their result and exclude them on grounds of probability.

If we accept this argument the concentration of the lighter isotope of carbon ^{12}C relative to ^{13}C found within these inclusions may be taken as strong evidence for the existence of microbial life on Earth before 4.1 Gyr, during the epoch of comet and asteroid impacts (the Hadean epoch). Based on orthodox views we now require that an essentially instantaneous transformation of nonliving organic matter to bacterial life takes place. Without constraints of orthodoxy a more plausible option is that fully developed microorganisms arrived at the Earth via impacting comets and these became carbonized and trapped within condensing mineral grain conglomerates (Fig. 1.7). The scientific consensus while moving away from an organic soup generated in situ on Earth to one supplied from space, still clings tenaciously to a preferred origin of life on the Earth.

9. Concluding remarks

A wide range of arguments that can be used in defense of panspermia have been discussed in detail elsewhere (Wickramasinghe et al., 2015). By way of concluding this article we refer to two recent discoveries that corroborate the earlier arguments of this paper. The first is the discovery of a microorganism at a depth of 2.8 km in a South African gold mine that derives its energy from radiolysis induced by particles emitted from U, Th, and K (Atri, 2016). *Desulforudis audaxviator* is a microbe that would not only survive interstellar transits but would thrive on energy derived from galactic cosmic rays for instance in the outer layers of free floating planets (Wickramasinghe et al., 2012).

References

Allen, D.A., Wickramasinghe, D.T., 1981. Nature 294, 239.

Altwegg, K., et al., 2016. Sci. A 2, 5.

Atri, D., 2016. J. R. Soc. Interfac. 13, 20160459.

Bell, E.A., Boehnke, P., Harrison, T.M., Mao, W.L., 2015. Proc. Natl. Acad. Sci. U. S. A. 112, 14518.

Bieler, A., et al., 2015. Nature 526, 678.

Biver, N., et al., 2015. Sci. A 1, e150086.

Brand, B.R., Bernard-Salas, J., Spoon, H.W.W., et al., 2006. Astrophys. J. 653, 1129−1144.

Capaccione, F., et al., 2015. Science 347, A0628.

Catzetti, D., Armus, L., Bohlin, R.C., Kinney, A.L., Koornneef, J., Storchi-Bergmann, T., 2000. ApJ 533, 682.

Chhowalla, M., Wang, H., Sano, N., et al., 2003. Phys. Rev. Lett. 90, 155504.

Deamer, D., 2012. First Life. Univ. California Press, Oakland.

Draine, B.T., 2003. ARA&A 41, 241.

Fagerbakke, K.M., Heldal, M., Norland, S., 1996. Aquat. Microb. Ecol. 10, 15−27.

Fitzpatrick, E.L., 1986. AJNR 92, 1068.

Hoyle, F., Wickramasinghe, N.C., 1962. MNRAS 124, 417, 41, 693.

Hoyle, F., Wickramasinghe, N.C., 1969. Nature 223, 450.

Hoyle, F., Wickramasinghe, N.C., 1977. Nature 270, 701.

Hoyle, F., Wickramasinghe, N.C., 1978. Lifecloud: The Origin of Life in the Galaxy. Dent, London.

Hoyle, F., Wickramasinghe, N.C., 1981. In: Ponnamperuma, C. (Ed.), Comets and the Origin of Life. Reidel, Dordrecht.

Hoyle, F., Wickramasinghe, N.C., 1991. The Theory of Cosmic Grains. Springer, Heidelberg.

Hoyle, F., Wickramasinghe, N.C., 2000. Astronomical Origins of Life: Steps Towards Panspermia. Kluwer, Dordrecht.

Hoyle, F., Wickramasinghe, N.C., Al-Mufti, S., Olavesen, A.H., 1982a. Ap&SS 81, 489.

Hoyle, F., Wickramasinghe, N.C., Al-Mufti, S., Olavesen, A.H., Wickramasinghe, D.T., 1982b. Ap&SS 83, 405.

Hoyle, F., Wickramasinghe, N.C., Al-Mufti, S., Olavesen, A.H., 1999. Ap&SS 268, 379.

Kumar, S.A., Wickramasinghe, N.C., Louis, G., 2019. Adv. Astrophys. 4 (2), 72.

Lynds, B.T., Wickramasinghe, N.C., 1968. ARA&A 6, 215.

Motta, V., Mediavilla, E., Munoz, J.A., et al., 2002. Astrophys. J. 574, 719.

Noll, S., et al., 2009. A&A 499, 69.

Posberg, F., Khawaja, N., Abel, B., et al., 2018. Nature 558, 564.

Prevot, M.L., Lequeux, J., Prevot, L., Maurice, E., Rocca-Volmerange, B., 1984. A&A 132, 389.

Rauf, K., Wickramasinghe, C., 2010. Int. J. Astrobiol. 9 (1), 29.

Scoville, N., Faisst, A., Capak, P., Kakazu, Y., Li, G., Steinhardt, C., 2015. ApJ 800, 108.

Seaton, M.J., 1979. MNRAS 187, 73.

Spencer, J.R., Nimmo, F., 2013. Annu. Rev. Earth Planet Sci. 41, 693.

Spitale, J.N., Hurford, T.A., Rhoden, A.R., et al., 2015. Nature 521, 57.

Taft, E.A., Phillipp, H.R., 1965. Phys. Rev. 138A, 197.

Thomas, P.C., Tajeddine, P., Tiscereno, M.S., et al., September 11, 2015. Icarus on Line.

van de Hulst, H.C., 1957. Light Scattering by Small Particles. Wiley, New York.

Wallis, M.K., Wickramasinghe, N.C., 2015. Astrobiol. Outreach 3, 12.

Wallis, M.K., 1980. Nature 284, 431.

Wickramasinghe, D.T., Allen, D.A., 1980. Nature 287, 518.

Wickramasinghe, D.T., Allen, D.A., 1986. Nature 323, 44.

Wickramasinghe, N.C., Guillaume, C., 1965. Nature 207, 366.

Wickramasinghe, N.C., Tokoro, G., 2014. J. Astrobiol. Outreach 2, 133.

Wickramasinghe, D.T., Hoyle, F., Wickramasinghe, N.C., Al-Mufti, S., 1986. EM&P 36, 295.

Wickramasinghe, N.C., Wallis, J., Wallis, D.H., et al., 2012. Astrophys. Space Sci. 341, 295.

Wickramasinghe, N.C., Wainwright, M., Smith, W.E., Tokoro, G., Al-Mufti, S., Wallis, M.K., 2015. Astrobiol. Outreach 3, 126.

Wickramasinghe, N.C., 1973. Interstellar Grains.

Yabushita, S., 1993. MNRAS 260, 819.

UV astronomy and the investigation of the origin of life

Ana I. Gómez de Castro[1,2], Ada Canet[1,2]

[1]*Joint Center for Ultraviolet Astronomy (JCUVA), Universidad Complutense de Madrid, Madrid, Spain;* [2]*U.D. Astronomia y Geodesia, Facultad de CC Matemáticas, Universidad Complutense de Madrid, Madrid, Spain*

1. Introduction

The scientific investigation about the origin of life started after the experiments by Oparin, Miller, and other pioneers showing that under certain physical conditions amino acids may naturally emerge on Earth, setting the first steps to the formation of proteins, RNA, and hopefully life. The understanding that this process can apply to many astronomical bodies (planets, moons, asteroids) orbiting around a fair fraction of the stars in the Universe led to the beginning of astrobiology as a science.

During the last two decades, laboratory experiments have proven that ultraviolet (UV) irradiation of cometary-like ice results in the formation of all the reference nucleobases for RNA and DNA: uracil, cytosine, thymine, adenine, and guanine. In a seminal work, Muñoz-Caro et al. (2002) and Bernstein et al. (2002) showed that small amino acids, such as glycine and alanine form after the UV irradiation of dirty, cometary-like, ices. Later on, further experiments showed that UV irradiation of N-heterocycles pyrimidine and purine results in the production of the nucleobases in terrestrial DNA (Nuevo et al., 2014; Materese et al., 2017). In the last few years, the cycle has been closed with a set of experiments producing ribose and 2-deoxyribose (Nuevo et al., 2018), and pyrimidine and purine (Oba et al., 2019), again, through the UV photo processing of interstellar-like ices.

This research argues for a space origin of the prebiotic building-blocks of life and sets the base for the *soft panspermia* theory. The evidence suggests that first life appeared on Earth earlier than 3.7 Gyr ago (Nutman et al., 2016) during a period of heavy meteoric bombardment. Therefore, the contribution of space generated nucleobases to the Earth's budget cannot be neglected; space bodies certainly delivered their enriched content into planets. Successive hydration and dehydration cycles in warm little ponds (WLP) have been shown to compete well with hydrothermal vents in the polymerization of nucleotides and the emergence of long RNA chains (Pearce et al., 2017).

The astronomical community is devoting an enormous effort to detect environments where life may have emerged in space. The contribution of astronomy to the investigation of the *origin of life* concentrates on few topics: understanding the key processes behind the chemical enrichment of the Universe, investigating the formation of planets, and searching for signatures of life (see, i.e., Hanslmeier et al., 2012, Kitadai and Maruyama, 2018; Camprubí et al., 2019). About half of the portfolio of the large astronomical facilities contains technical details on the requirements to be optimized for this research. There are many excellent reviews on the role that large astronomical facilities will play in this investigation (Des Marais et al., 2002; Seager 2014; Cockell et al., 2016; Grenfell 2017; Kaltenegger 2017; Walker et al., 2018; Schwieterman et al., 2018; Kiang et al., 2018; Catling et al., 2018). This contribution concentrates on the role of observatories working at UV wavelengths on this endeavor. The UV spectrum carries crucial information about astronomical environments of extraordinary interest for astrobiology such as the young planetary systems. Moreover, the actual knowledge of the UV radiation field is instrumental for the interpretation of data (obtained at any wavelength) related with magnetic activity, photon driven heating, and astrochemistry.

In this chapter, we summarize the main effects of UV radiation on matter within the context of astrobiological research. In subsequent sections, we address the UV observations of the evolution of planetary systems (Section 3) and exoplanets (Section 4), including the research to detect biosignatures. For completeness, a short section on the chemical enrichment of galaxies and the associated implications for the origin of life studies is included (Section 5). Remote detection of amino acids and other complex molecules involved in the formation of RNA is the holy grail of astrobiology. Radioastronomical observatories have detected some complex molecules and the high sensitivity of the Instituto de RadioAstronomia Milimetrica (IRAM) 30-m telescope at Pico Veleta (Spain) and the Atacama Large Millimeter Array (ALMA) are revolutionizing this field of research with the detection of new molecules such as urea (see, i.e., Jiménez-Serra et al., 2020). However, the chirality of some of the most abundant amino acids, such as the alanine, may open a brand-new path to the remote sensing of life-related molecules through the use of UV spectropolarimetric techniques (Gómez de Castro & Isidro-Gómez, 2021). Section 6 summarizes the requirements in terms of polarimetric accuracy and sensitivity for detecting alanine in Solar System bodies. The work concludes with a short appraisal of the needs for UV observatories and instrumentation (Section 7).

2. The interaction of UV radiation and matter: implications for astrobiology

There are extensive treaties describing the atomic and molecular processes behind the interaction of radiation with matter in the many astrophysical environments (see, i.e., Spitzer 1978; Osterbrock and Ferland 2006; Rybicki and Lightman 2008).

UV photons with energies between 4 and 14 eV are able to photoionize and photo-excite the most abundant species in astronomical environments (see Fig. 2.1), some of the most abundant molecules (H_2, HD, CO, OH, H_2O, S_2), and drive important photochemical reactions. The electronic bands of many molecules relevant to *origin-of-life* studies are in the UV. O_2 and especially its photochemical byproduct O_3, possess strong absorption bands in the UV range. Common molecules detected in the UV spectra of cometary tails are: OH, CO, C_2, CS, CO_2^+, CO^+, S_2 (Weaver et al., 1981, see also Kim and Hearn 1991, and references therein for an analysis on SO and S_2O presence).

The most abundant molecule in the Universe is H_2 but its rovibrational infrared transitions are dipole forbidden because of the homonuclear nature of the molecule; this makes H_2 difficult to detect at infrared wavelengths (Pascucci et al., 2006) however, the electronic transitions are pumped by the stellar Lyα photons enabling the detection of H_2, at UV wavelenghts, in the inner border of the protoplanetary disks even at the late stages of the PMS evolution (France et al., 2012b).

Polycyclic Aromatic Hydrocarbons (PAHs) are basic ingredients for organic chemistry in astronomical environments. Their resonant absorption of photons with wavelengths ~ 220 nm contributes to the prominent bump observed in the interstellar extinction curve at UV wavelengths (Joblin et al., 1992); other possible contributors are graphite particles (Stecher and Donn 1965), fullerenes (Iglesias-Groth 2004), coal-like materials (Papoular et al., 1996), diamonds (Saslaw and Gastad 1969), and many other carbon-based particles (see Malloci et al., 2008 for a description).

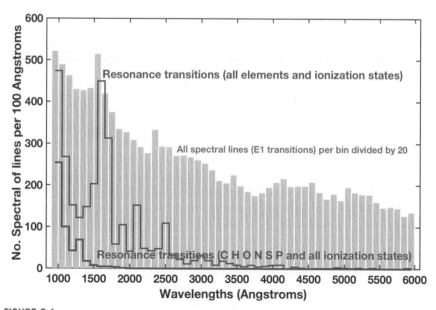

FIGURE 2.1

Density of spectral lines per 100 Angstroms from the UV to the optical range. Resonance transitions of all elements and resonance transitions of the most relevant elements for astrobiology (C, H, N, O, S and P) are plotted in colors. The background grey bars indicate the distribution of all E1 transitions, of all elements for comparison. Data have been taken from the Atomic Line List maintained by Peter van Hoof (https://www.pa.uky.edu/~peter/newpage/).

The degree of hydration of the lunar soils is measured in the 160−180 nm range where the strong electronic band of H_2O is detected (Hendrix et al., 2012).

The strong interaction between UV photons and hydrogen, the most abundant element in the universe, has two important and well-known astrophysical effects: the formation of H II regions caused by the photoionization of hydrogen and the photodissociation of the H_2 molecule leading to the formation of photodissociation regions (PDR) in the outskirts of molecular clouds or in the atmosphere of protostellar disks.

In astrochemistry, the low space densities prevent the charged particles from interacting efficiently to form large molecules but, on the icy surfaces of interstellar dust grains, comets, asteroids or planets, UV photons provide electrical charge, and hence mobility, directivity, and energy for chemical reactions (Tielens and Allamandola 1987; Grim and Greenberg, 1987; Gerakines et al., 1996; Tielens, 2020).

Within the context of the investigation of the origin of life on Earth, there are two key processes to be considered: [1] the survival of micro-organisms under UV irradiation, and [2] the role of stellar UV radiation as a driver of astrochemical processes, especially on ices. Moreover, photons with energies above 6 eV generate a photoelectric flow from the dust grains (Spitzer 1978; Weintgarner & Draine 2001) that heats the surrounding gas, affecting the chemistry, and that depends on the spectral energy distribution of the radiation. In graphite, the ejected photoelectrons lead to the formation of faults in the layered structure of graphite leading to a phase transition to diamond. This reduces the amount of carbon atoms available for the chemical processes.

2.1 UV radiation and the survival of lifeforms

Massive UV radiation is lethal to most forms of life and affects both the growth of living forms in planets and their spread in space. UV irradiation can denature the DNA of microorganisms, causing death or inactivation resulting from strand breakage, as well as the generation of mutagenic photoproducts that inhibit the transcription/replication of chromosomes (Singh et al., 2010). The UV absorbance of DNA peaks around 260 nm; at lower and higher wavelengths, the absorbance decreases. Below 230 nm, the absorbance increases again (see Fig. 2.2).

The effectiveness of water to palliate this damage has been extensively studied in the context of UV disinfection technology (Hijnen et al., 2006). For any given microorganism, this effectiveness depends not only on the total UV flux (or fluence) but also on the energy distribution of the radiation field. The most UV resistant organisms are viruses and bacterial spores.

Since the advent of the space era, a number of life forms (from microorganisms to small life forms) have been submitted to space conditions of gravity, cosmic radiation, and vacuum. The *Bacillus subtilis* is a well-known example; it can survive for years in low Earth orbit provided it is protected against the UV radiation in space. Exobiology experiments have obtained similar results for other organisms including viruses, bacteria, fungi, and nematodes (Nicholson et al., 2000; Rothschild and

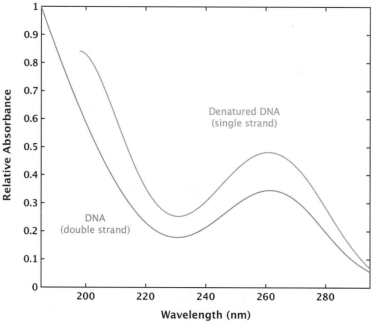

FIGURE 2.2

UV absorbance of native and heat-denaturated *Escherichia. coli* DNA. The purine and pyrimidine bases in DNA strongly absorb UV radiation. Upon denaturation, the double stranded DNA becomes single and the bases are exposed hence increasing the absorbance.

Mancinelli, 2001; Novikov et al., 2011). In this context, the transportation of microbes between planets seems feasible provided the microorganisms are encapsulated within meteoroids which supply protection against solar UV radiation, atmospheric entry heating and the subsequent shock. Calculations show that if life existed on Mars, transport to Earth was very probable given the number of Martian meteorites that have fallen on Earth (Mileikowsky et al., 2000). Likewise, human expansion in space will inevitably result in the spread of Earth-based microorganisms into Solar System bodies (even with the sterilization policies implemented by the main space agencies).

At the start of the Proterozoic era, 2.4 Gyr ago, the growth of cyanobacteria in the Earth's oceans led to an increase of the oxygen abundance in the atmosphere and the formation of the ozone layer that blocked efficiently the solar UV radiation. However, the discovery of stromatolites and microfossils in old sedimentary rock formations indicates that life started much earlier on Earth, ∼3.7 Gyr ago (Walsh and Lowe, 1985; Ohtomo et al., 2014), at a time when the harmful, 110−290 nm wavelength, solar UV radiation was reaching the surface. Early living forms may have grown in the depth of the oceans protected by water from UV radiation however, they also seem to have flourished on the sediments at the bottom of shallow waters, in WLP exposed to the solar UV flux, only slightly attenuated by the overlaying water. It seems that ferric iron, abundant in these sediments, could have protected the earliest forms of life because of its strong absorbance of UV radiation (Olson and Pierce, 1986). Dissolved and particulate matter suspended in the water may also have reduced the UV flux in the Archean ocean however, calculations show that the UV radiation would penetrate sufficiently deep in the water to produce significant DNA damage through the mixed layer (Cockell, 2000) hindering the growth of early cyanobacteria.

Modern cyanobacteria have developed mechanisms to counteract the much softer UV radiation field getting through the ozone layer (O'Brian and Houghton, 1982, Owttrim and Coleman, 1989, Mittler and Tel-Or, 1991, Ehling-Shultz et al., 1997, Mlozewska et al., 2018). However, some strains of cyanobacteria are too small to accommodate enough sunscreen molecules for effective protection against a sudden increase of the UV radiation (García-Pichel, 1994) and solar flares and super-flares may affect their survival and that of other Earth's microorganisms (see Chapter 3 in this volume).

Interestingly, some current terrestrial living forms have developed protective skills against the hard UV radiation in outer space. Tested examples are bacteria (Hörneck et al., 2001a,b), dry lichens (Sancho et al., 2007), and tardigrades (Jönson et al., 2008). *Rhizocarpon geographicum* and *Xanthoria elegans* were exposed to UV space conditions for 16 days, as part of the 2005 LICHENS experiment. Upon return to Earth, they recovered nearly the same photosynthetic activity and showed no significant structural changes; the accumulation in their upper cortex of compounds able to filter UV radiation seems to have provided adequate protection to the lichenized fungal and algal cells (Solhaug et al., 2003). Tardigrades are among the most desiccation and radiation tolerant animals, including exposure to extreme levels of ionizing radiation in the laboratory however, only few specimens of *M. tardigradum* survived exposure to the full spectrum of the solar UV radiation in low Earth orbit (Jönsson et al., 2008). In summary, there are living forms on Earth able to survive direct exposure to space UV radiation.

2.2 UV radiation driven organic chemistry

From the astrophysical point of view, organic materials are disequilibrium chemical species produced only in certain environments that require low-temperature, partial ionization, and UV-irradiation such as the interstellar medium or the proto planetary disks. The most interesting laboratories for astrobiological purposes are the disks around young stars and their later evolution, the icy bodies in planetary systems.

The physical conditions of protoplanetary disks cover a broad range of densities ($1\text{-}10^{13}$ cm^{-3}), temperatures ($10\text{-}2{,}000$ K) and irradiation levels. They depend on the distance to the star, the vertical scale height within the disk, the stellar and environmental radiation field and the evolutionary state of the disk which is basically measured through the gas-to-dust ratio (see Section 3). Molecular hydrogen is the dominant component and it does not freeze out onto dust grains. The second most abundant gaseous component is CO and its abundance is significantly lower (CO/ $H_2 \sim 10^{-4}$) hence, the evolution of the disks is controlled by the H_2 molecule.

The H_2 molecule is photodissociated through the absorption of UV photons in the Lyman and Werner bands (912−1100 Å) though H_2 self-shields efficiently and only about 15% of the UV photons actually lead to H_2 photodissociation. This process drives the formation of a PDR at the surface of protoplanetary disks that acts as a transition layer between the tenuous, warm, ionized atmosphere, and the dense, cold molecular gas in the interior. The transition between the atomic and the molecular hydrogen layer is sharp and propagates into the disk interior as the warm material evaporates; the neutral gas is ionized by the Lyman continuum radiation that heats it to some 10,000 K surpassing the escape velocity from the disk surface and generating feeble photoevaporative thermal flows (Hollenbach et al., 2000).

The vertical structure of the disk is determined by the environmental UV and X-ray radiation field. According to its penetration depth, the disk is divided in three basic layers: the dense midplane, opaque to both X-ray and UV radiation and primarily ionized by cosmic ray particles, the disk atmosphere, which is transparent to the stellar UV radiation and an intermediate layer opaque to UV photons but transparent to X-ray (see Fig. 2.3).

The gas phase chemistry in the disk atmosphere is mainly driven by the intense UV radiation fields and the subsequent ionization and/or destruction of small molecules. Large molecules can withstand the high radiation fields; PAHs and their derivatives, as well as aliphatic components, are the main organic molecular species (Tielens, 2020). UV photons acting on small PAH molecules create radical cations and neutral radicals facilitating PAH's interaction and growth (Zhen et al., 2018).

Another well-known effect is the impact of the UV radiation field on the CO photodissociation rate. A reduction in the atomic carbon supply leads to smaller abundances of CN and C_2N radicals (van Zadelhoff et al., 2003; Bergin et al., 2003), which is relevant to the formation of cyanopolynes and amino acids (Winstanley and Nejad, 1996; Fukuzawa et al., 1998; Agúndez et al., 2017). Gas phase reactions and their rate coefficients are available in the UMIST RATE12 database (McElroy et al., 2013).

The denser parts of the disks are not completely devoid of ionizing radiation. Cosmic rays are able to penetrate into the disk interior and their interaction with matter generates energetic electrons susceptible to excite H_2 molecules that subsequently, release UV photons from their deexcitation (Prasad and Tarafdar, 1986). Photoionization and photodissociation produce ions and radicals from small species (H_2, CO, H_2O, HCO, HCN, H_2CO) that have larger cross-sections for chemical reactions than neutrals and usually react faster (Herbst, 1988). Molecules very relevant for the disk organic chemistry such as CH_2O or CH_3OH may be synthesized in this manner (Wirstrom et al., 2011).

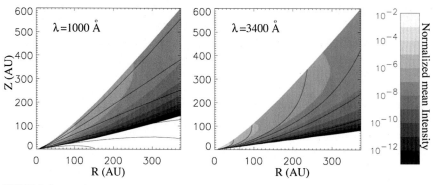

FIGURE 2.3

Penetration of the stellar UV radiation field into a protoplanetary disk. Both the galactic background UV radiation field and the stellar radiation are considered for the calculation; the star is located at the origin. The mean intensities are normalized to the value in the innermost cell. Overplotted are the H_2 number density contours (from surface to midplane: 5×10^4, 2×10^5, 10^6, 10^7, 10^8, 10^9) [cm^{-3}] on the left and the dust temperature contours (upper left to lower right: 90, 70, 50, 40, 30, 20) [K] on the right.

After van Zadelhoff et al. (2003).

Dust grains are key ingredients in the disk chemistry; they provide the matrix for large molecules to grow. At the low temperatures of the disk midplane large molecules are stationary on the grain surface and only grow by reaction with mobile species like hydrogen. However, as the temperature rises, diffusion and thermal evaporation produce desorption of the light species such as hydrogen, thus braking the hydrogenation driven chemistry on the surface. A further increase, results on heavier compounds leaving the grain surface. This desorption process depends exponentially on the grain temperature, as well as on the gas phase abundances and the chemistry. As a result, the chemical evolution of protoplanetary disks is very sensitive to the processing of the grain mantles (Hasegawa et al. 1992, Hasegawa and Herbst, 1993).

Moreover, particles are not stationary in protoplanetary disks. Disk dynamics results in a mixing of materials and a cycling between zones having different temperatures and radiation fluxes; thus, materials experience very different thermal and radiation processes. For instance, the vertical motions of particles within the disks increase their UV radiation dose and leads to multiple condensation-photoevaporation cycles; icy particles from the midplane move to upper layers of the disk where there are enough UV photons for the photolysis of the ices to occur to later on, descend again, and accumulate more ices.

2.2.1 Chemistry of astronomical ices
Gas phase chemistry is rather inefficient to produce complex molecules compared with reactions in the icy mantles of the dust grains. On the surface of cold grains,

species are closer to each other and the ice matrix provides shelter and stability to the newly formed molecules, also increasing the survival time of some ions and radicals. Processing of ices with energetic photons and/or particles is an important source for organic molecules in Solar System bodies and in other planetary systems setting the astrochemical basis for panspermia (see Chapter 1 and references therein).

Laboratory experiments show that UV irradiation of even the simplest compounds leads to the formation of new molecular species, including amino acids, sugar derivatives and other organic molecules of interest for astrobiological research (Nuevo et al., 2010; de Marcellus et al., 2011). The type of materials produced is rather insensitive to the precise composition of ices, provided it includes usable sources of the fundamental molecular compounds: C, H, O, and N. For example, C can be supplied in the form of CH_3OH, CH_4, CO, CO_2 or any combination of them resulting, in all cases, in the formation of amino acids (Muñoz-Caro et al., 2002; Bernstein et al., 2002; Nuevo et al., 2008) though their relative abundance may vary. Similar products are obtained from H_2O ice at temperatures from 10 to 150 K (Bernstein et al., 1995). Irradiation of a mixture containing the simplest PAH, naphthalene, resulted in a higher abundance of amino acids containing aromatic rings (Chen et al., 2008). The process is also rather insensitive to the source of ionization either be high energy protons (i.e., cosmic rays) or UV radiation (Gerakines et al. 2001, 2004).

The most abundant amino acids produced by UV photoprocessing of ices are glycine and alanine (Muñoz-Caro et al., 2002; Bernstein et al., 2002) and, in general, the abundance of amino acids decreases with their molecular mass.

The set of amino acids produced in these experiments and their relative abundances are rather similar to those observed in meteorites and comets (Shock and Schulte, 1990; Altwegg et al., 2016) however, the production of levo and dextro isomers of the same chiral molecule is well balanced in laboratory experiments while enantiomeric excesses of 1%−10% are typically observed in meteorites (Pizzarello and Cronin, 2000) and may reach as high as a 60% (Pizzarello et al., 2012). This unbalance can be induced by the irradiation of the ice with polarized UV radiation as proved by laboratory experiments; a nonracemic mixture of alanine with enantiomeric excess <4.2% has been produced in the laboratory in this manner (Meinert et al., 2014).

Another family of organics relevant to the investigation of the origin of live are sugars and sugar derivatives. Ribose is a building block for RNA and 2-deoxyribose for DNA. Moreover, they take part on the cell walls and provide a reservoir for energy storage. Ribose and 2-deoxyribose are produced after UV irradiation of simple ice mixtures containing H_2O and CH_3OH (Nuevo et al., 2018). Recent experiments have shown that several pyrimidine- and purine-based nucleobases can be formed from the UV irradiation of simple H_2O, CO, NH_3, CH_3OH ice mixtures at 10 K (Oba et al., 2019). UV irradiation of N-heterocycles pyrimidine and purine in astrophysical ices with a high abundance of H_2O but also containing NH_3 and CH_4 results in the production of all the reference nucleobases for RNA and DNA; uracil, cytosine, thymine, adenine and guanine have been produced in this manner (Nuevo et al., 2014; Materese et al., 2017).

In summary, the last decade of astrobiological research has shown that the UV radiation of astrophysical ices under space conditions results in the formation of the building blocks needed to produce RNA and DNA.

Interestingly, the set of amino acids produced by the irradiation of astronomical ices differs from the results of classical experiments where amino acids are formed under electrical discharges in a reducing atmosphere (Miller, 1953; Miller and Urey 1959; Oró 1967; Cleaves et al., 2008; Parker et al., 2011). In Table 2.1, the proteinogenic amino acids formed through both routes are shown for some representative experiments.

Table 2.1 Proteinogenic amino acids produced in some of the "origins of life" experiments.

Proteinogenic Aminoacids	Laboratory experiments simulating			
	Early Earth conditions		Space ice & comets	
	Miller (1953)	Parker et al. (2011)	Muñoz-Caro et al. (2002)	Chen et al. (2008)
Glycine	Produced	Produced	Produced	Produced
Alanine	Produced	Produced	Produced	Produced
Aspartic acid	Produced	Produced	Produced	Produced
Serine		Produced	Produced	Produced
Valine		Produced	Produced	
Glutamic acid		Produced		Produced
Phenylalanine				Produced
Methionine		Produced		
Isoleucine		Produced		Produced
Leucine		Produced		Produced
Cysteine				
Histidine				Produced
Lysine				
Asparagine				
Pyrrolysine			Produced	
Proline				
Glutamine				
Arginine				Produced
Threonine				
Selenocysteine				
Tryptophan				
Tyrosine				Produced

2.3 **UV spectral energy distribution and the photoelectric yield**

In diffuse environments such as the envelopes of the molecular clouds, the atmosphere of young planetary disks or the interplanetary medium, UV photons are absorbed by materials in the dust grains (graphites, silicates and ferrites, mainly) yielding to the emission of photoelectrons that heat and ionize the environment and accelerate some chemical processes. Only UV radiation, with energy above ~ 6 eV, is able to produce a significant photoelectric yield.

The heating rate depends on the density and energy of the ejected photoelectrons which, in turn, depends on the dust grains size (Watson, 1972), composition and charge state as well as on the spectral energy distribution (SED) of the radiation feeding the photoelectric flow (i.e. Pedersen and Gómez de Castro, 2011). The photoelectric yields are displayed in Fig. 2.4 (after Weingartner and Draine, 2001) for silicates and carbonaceous particles. Photons with energies above some 6 eV produce a significant yield from carbonaceous particles while the threshold goes up to 8 eV for silicates.

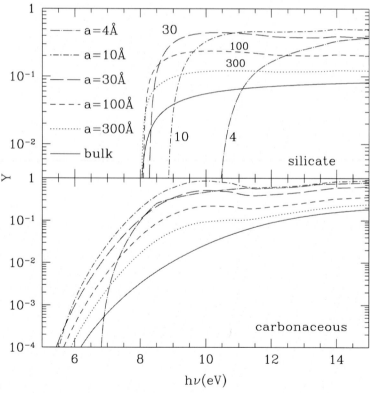

FIGURE 2.4

Photoelectric yield of silicate and carbonaceous particles in space. The various curves correspond to different particle size. In general, the smallest the size the larger the yield because of the finite electron escape length.

After Weintgarner & Draine (2001).

The effectiveness of the process depends on the kinetic energy acquired by the photoelectrons to overcome the Coloumb field of the charged dust grain. Hence, there is a strong dependency between the final charge of the grains and the SED of the UV spectrum irradiating them. This effect shows in the temporal evolution of the charge of the grains (or charging profile) when submitted to a specific UV spectrum, as shown in Fig. 2.5. Two very different spectra are selected to illustrate this process: the very hot, but soft SED of an O-type star and the emission lines spectrum of a pre-main sequence (PMS) solar-like star. Above 6 eV, PMS stars only radiate significantly in very narrow bands associated to the main spectral lines at: ~ 8 eV (the He II at 164 nm and CIV at 1550 nm lines), 8.9 eV (Si IV), 9.3 eV (CII), 9.5 eV (OI) and, of course, the $Ly\alpha$ hydrogen line at 10.2 eV. This is the radiation affecting the dust grains in young planetary disks. Dust in the HII regions and the bright reflection nebula are however, charged by the smooth UV spectrum of hot stars. Some examples are the B33 (Horsehead) nebula irradiated by a cluster of early B-type late O-type stars (Caballero, 2007) and the Rosette nebula by the cluster NGC 2244 of B8-9 type stars and 7 O4−O9 stars (Martins et al., 2012). As shown in the bottom panel of Fig. 2.6, dust charging profiles are step-wise in young planetary disks instead of the smooth rising curves from HII regions which are, otherwise consistent with the experimental observation (Sickafoose et al., 2000).

As interstellar and interplanetary dust grains are not in isolation, the final grain charge will also depend on the temperature of the surrounding gas that acts as a regulator since dust grain's charge is modified in the gas-dust collisions (Spitzer, 1978).

2.4 UV radiation driven diamondization of graphites

A major fraction of the heavy atoms in space are locked up in the dust grains that aggregate into larger structures to build comets, minor bodies and planets in the circumstellar disks around young stars. The bulk of the grain mass is provided by the most abundant elements namely, C, N, O, Mg, Si and Fe. The expected most abundant species in young planetary disks are gathered in Table 2.2.

Carbon is a fundamental element in organic chemistry and comes mainly in several refractory organic compounds in the dust grains (Jenkins et al., 1983, see also Table 2.2). The preferred allotropic form of carbon is graphite (Draine, 2016) where carbon atoms are arranged in sheets, weakly bound together by van der Waals forces. The interlayer separation is ~ 3.4 Å and within each sheet, the atoms are disposed in a honeycomb lattice, each linked to the three neighbours through covalent sp^2 bonds.

Interestingly, nanodiamons are the third most abundant specie in Solar System meteorites (Huss et al., 2003). Carbon atoms in nanodiamonds are strongly bonded with sp^3 bonds in a cubic network and thus, removed from the budget available for the chemical reactions in the dust grains. Strong bonding is also characteristic of much less abundant species such as fullerenes and buckyonions, expected to exist in the interstellar and interplanetary environment (Iglesias-Groth, 2004).

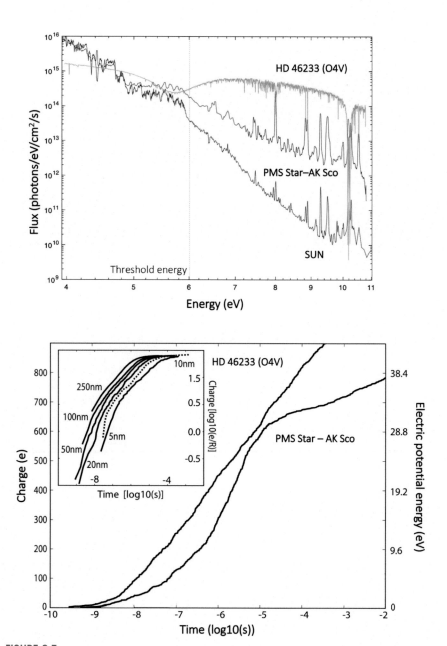

FIGURE 2.5

Top: STIS spectra of the Sun (in red, from Meftah et al., 2020), the PMS binary AK Sco (blue, from Gómez de Castro et al., 2020b) and the O-type star HD46223 (yellow, data from the Hubble archive). AK Sco and HD46223 fluxes are scaled so the integrated flux is the same as the solar flux received from the Earth at 1 AU in the STIS wavelength range. **Bottom:** Charging profile of a silicate dust grain (a = 30 nm) submitted to the radiation fields of an O-type and a PMS star (Pedersen & Gómez de Castro, 2011).

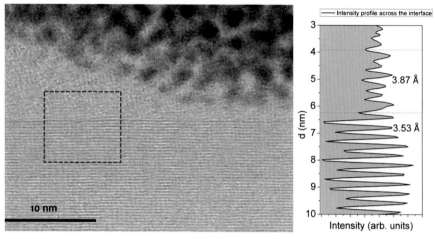

FIGURE 2.6

Annular bright field (ABF) images of a Highly Ordered Pyrolytic Graphite (HOPG) sample after irradiation with a Deuterium UV lamp. The high-resolution scanning transmission electron microscopy (STEM) images resolve the separation between the individual graphene layers in the HOPG. The image evidences the dislocation of the superficial graphitic layers after irradiation and the formation of bumps. Electron Energy Loss spectroscopy (EELs) of the bumps show that the sp^2 to sp^3 hybridation transition is a surface effect that is maximum at the irradiated surface. High sensitivity X-ray diffraction experiments confirm the formation of diamond within the bumps. On the right, the separation between the graphene layers is plotted. There is a transitional area where separation between the layers increases from 3.53 Å to 3.87 Å.

After Gómez de Castro et al., (2020a).

Table 2.2 Expected most abundant dust grain species in young planetary disks.[a]

Specie	Bulk density (g/cm^3)	Mass fraction
Water ice (H$_2$O)	0.92	5.55×10^{-3}
Refractory organic (graphite, PAHs, CHON, nanodiamonds …)	1.5	3.53×10^{-3}
Olivine (Mg,Fe)$_2$SiO$_4$	3.4	2.51×10^{-3}
Troilite (FeS)	4.83	7.68×10^{-4}
Orthopiroxene (Mg,Fe)$_2$Si$_2$O$_6$, (Mg,Fe)SiO$_3$	3.49	7.33×10^{-4}
Volatile organic (CO, COH$_2$, CH$_n$)	1.0	6.02×10^{-4}
Metallic iron	7.87	1.26×10^{-4}

[a] after Table 2.2 in Pollack et al. (1994).

Recently, it has been shown that the UV irradiation of graphite leads to the formation of nanodiamonds (Gómez de Castro et al., 2021). Photons with energies above 6 eV (UV photons) are able to release the loose electrons in the graphite's π orbitals from the graphitic matrix, creating holes that render unstable the graphitic layered structure and lead to the formation of a cubic network: diamond. As UV radiation is heavily absorbed by both diamond and graphite, this is a surface process, as illustrated in Fig. 2.4, and results in the formation of sparse nanodiamonds on the surface of the interstellar and interplanetary graphite grains (Gómez de Castro et al., 2021). These nanodiamonds are similar to those detected in meteorites (Abdu et al., 2018; Huss and Lewis, 1994) which leave their imprint in the extinction curve (Rai and Rastogi, 2010) and possibly in the anomalous microwave emission (Greaves et al., 2018). This process could also be responsible of the abundance of lonsdaleite in meteoritic samples.

The investigation of the Cañón del Diablo meteorite uncovered another sp^3 allotrope of carbon, the lonsdaleite, where carbon atoms are arranged in a hexagonal sp^3 crystalline structure (Frondel and Marvin, 1967; Hanneman et al., 1967; Lewis et al., 1987). Lonsdaleite is the less common of the two diamond polytypes and it is not detected free in nature. In meteoritic samples, it comes intermingled within the diamond network (Németh et al., 2014).

3. Formation of planetary systems

Planetary systems are end-products of star formation and, as such, their evolution is firmly tight to the physics of accretion, i.e., to the evolution of the circumstellar gas and dust under gravity, magnetic fields and the hard radiation from the star. Under the most common circumstances, accretion leads to the formation of disks or reservoirs where mass (and gravitational energy) is stored. Disk evolution is governed by accretion that is enabled by internal transport processes (Balbus and Hawley, 1991; Balbus and Papaloizou, 1999; Gammie, 2001) and gravitational waves (Lin and Papaloizou, 1996; Rice et al., 2005) while yielding powerful outflows (Goodson et al., 1997) and jets (see Gómez de Castro, 2013a and Hartmann et al., 2016 for reviews of accretion and outflows in the context of the solar-like PMS stars).

UV radiation abounds in these systems. It is produced in the accretion shocks where the gravitational energy of the infalling matter is finally released. It is also released in the dissipation of the magnetic energy built during the accretion process and the outflows thus generated. Moreover, solar-like PMS stars or T Tauri Stars (TTSs) are magnetically active and produce strong winds and outflows that also radiate in the UV. As a result, TTSs release ~ 50 times more energy in the UV range than their main sequence analogues.

This UV radiation has a profound effect in the evolution of the protoplanetary disks (Owen et al., 2012; Gorti et al., 2015). Together with X-rays (Ercolano et al., 2009), UV photons provide the energy to drive feeble photoevaporative flows from the surface of young planetary disks. Photoevaporation coupled with the dynamics of accretion creates a gap in the inner disk, at the radius where the photo-evaporative and the accretion rate become comparable (Clarke et al., 2001). The removal of H_2 from the disk stops accretion because the mechanisms enabling accretion operate mainly through the gaseous component.

The dynamics of the disk itself expands rapidly the gap and in less than 10 Myrs, the gas to dust mass fraction of the disk drops from 100 (typical interstellar value) to less than 4 in transitional disks (see i.e., Collins et al., 2009) and later on to values <<1 in debris disks (Roberge et al., 2006; Owen et al., 2012; Anderson et al., 2013), as illustrated in Fig. 2.7. Environmental radiation from nearby bright UV sources may also assist the process (Rosotti et al., 2017).

Three different radiation driven heating mechanisms (and radiation fields) are usually explored when considering disk photoevaporation:

- Far-Ultraviolet (FUV) radiation, with energies ranging from 6 to 13.6 eV, is capable of dissociating the H_2 molecule, and later, ionize and heat the atomic gas. It also generates a flux of photoelectrons from the dust grains that heat the gas through electron-particle collisions.
- Extreme-ultraviolet (EUV) radiation, with energies ranging from 13.6 eV to 0.1 keV, ionizes hydrogen.
- X-ray radiation, with energy above 0.1 keV up to 1−2 keV, produces K-shell ionization of abundant elements (O, C, Fe) resulting on a flux of energetic photoelectrons that collisionally excite and/or ionize the H, H_2 (Ercolano et al., 2009).

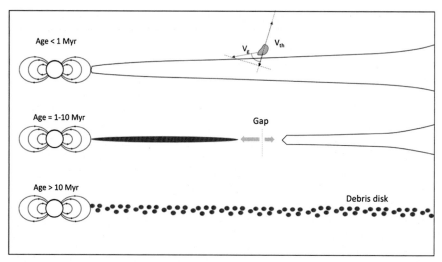

FIGURE 2.7

Evolution and dispersal of the stellar disks from accreting young stellar objects to debris disk.

The effect of EUV photons was the first to be explored (Hollenbach et al., 1994; Clarke et al., 2001) however, it leads to rather limited mass loss rates, even with the addition of some refinements (Alexander et al., 2014). FUV (Gorti and Hollenbach, 2009) and X-rays radiation fields (Owen et al., 2012) produce mass losses two orders of magnitude larger than the EUV flux. Normalized mass-loss profiles for the different photoevaporative radiation fields are shown in Fig. 2.8 for a $1M_\odot$ star.

The photoevaporative flow driven by the FUV radiation is mainly launched from the atomic layer. The depth of the FUV heated layer is equivalent to an N_H column density of $\sim 10^{21}-10^{23}\text{cm}^{-2}$ and depends primarily on the dust properties (and it is comparable to the depth of the X-ray-heated region). The temperature in the launching region varies significantly, ranging from $\sim 3{,}000K$ at the inner border of the disk (0.1 AU) to less than 100K at $\sim 100AU$, in a solar mass TTS. As shown in Fig. 2.8, the contribution of the FUV radiation to the mass-loss profile is very relevant at 5-10 AU and also, far from the star (radii > 100 AU) hence, it plays a major role in depleting the disk's gas reservoir and also contributes to the truncation of the disk at large radii.

Unfortunately, FUV-dominated photoevaporation rates are still uncertain because of the numerical complicacy of solving the radiative transfer and the dynamical problems simultaneously. Moreover, the uncertain abundance of PAHs in the gas phase (Gorti et al., 2015) has important thermal effects that may lead to order of magnitude variations in the calculations.

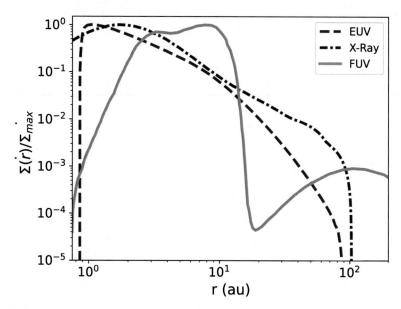

FIGURE 2.8

Normalized mass-loss profiles for different photoevaporative radiation fields around a $1M_\odot$ star (after Alexander et al., 2014). Radiation fields: EUV- (red, Font et al., 2004); X-ray- (blue; Owen et al., 2012); and FUV-dominated (green; Gorti et al., 2009).

Furthermore, FUV photons from the wind and the jet, may end being an important source of energy for photoevaporative flows if disk winds are thick enough to absorb a significant fraction of the stellar X-rays and UV radiation (Gómez de Castro and Ustamujic, 2015).

In addition to its relevance in disk evolution, observation of the UV radiation from PMS stars provides crucial information to study and understand the main processes leading to the formation of planetary systems, as it is shown below.

3.1 UV observation of the formation of planetary systems

In the current magnetospheric accretion paradigm (Romanova et al., 2012), the strong stellar magnetic fields truncate the inner disk at a few stellar radii (Donati and Landstreet, 2009). Gas flows from this truncation radius onto the star via the magnetic field lines forming accretion shocks at the impact point (Koenigl, 1991). The kinetic energy from the shock is released into heating at the stellar surface where the temperature reaches ~ 1 MK and strong UV and X-ray emission is produced (Lamzin, 1998; Gullbring et al., 2000). This magnetic core is embedded into an opaque dust cocoon during the first Myr of the evolution of a solar-like star. After, it becomes transparent to UV radiation. This is the relevant period to study the evolution from protoplanetary disk to planetary system.

The richness of the UV spectrum enables observing simultaneously the H_2 emission from the disk, the stellar wind, the bipolar jets, the photoevaporative wind from the disk, the stellar atmosphere (chromosphere, transition region, even corona through some very high ionization species such as Fe XII, Fe XVIII or Fe XXI), the magnetosphere and the accretion shocks. This is also the radiation that photoprocesses the disk materials and drives the organic chemistry on the icy coatings of dust grains. In Fig. 2.9, the UV spectrum of AK Sco is displayed and some of the main features are indicated.

3.1.1 Jets and outflows at UV wavelengths

Accretion onto magnetized sources drives the bipolar outflows that show as large-scale optical jets and molecular outflows carrying away the angular momentum excess of the accreting gas (see e.g., Pudritz et al., 2007). This mechanism taps the matter flow and controls the lifetime of the gaseous component of the disk through accretion and photoevaporation. At the innermost part of the disk, the star-disk interaction inflates the stellar magnetic field lines that extend up to several stellar radii (von Rekowski and Brandenburg, 2004), as shown in Fig. 2.10. A current layer is formed between the regions where the magnetic field is dominated either by the star or by the disk. It also marks the boundaries between the stellar magnetosphere, the stellar wind and the disk wind. Only at UV wavelengths all these components can be observed simultaneously.

There are three main outflows in the TTSs: stellar wind, disk wind and episodic mass ejections. The stellar wind is thin and so far, has not been detected. Its signature is feeble compared with dominant accretion tracers during the TTS phase; the only

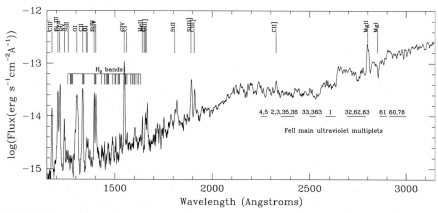

FIGURE 2.9

UV spectrum of AK Sco. The main spectral lines are indicated as well as the Fe II multiplets and the location of the H₂ molecular bands.

After Gómez de Castro et al., (2020).

possible evidence comes through the detection of shocks between the wind and remnant clouds of gas in the young planetary disks (Gómez de Castro, 2002). Disk winds and episodic ejections are best studied at the base of the bipolar jets. Episodic ejections are expected to be associated with recurrent phenomena such

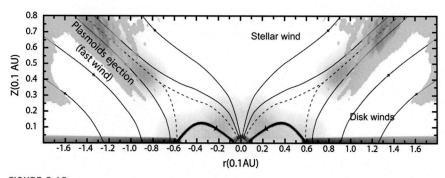

FIGURE 2.10

Expected location of the main components of the accretion engine; the interaction of the stellar magnetic field with the infalling plasma drives the outflow and the generation of reconnection layers (*dashed lines*). On top the C III] (191 nm) emissivity caused by the star-disk interaction according to MHD simulations is displayed in green-brown (see Gómez de Castro and von Rekowski, 2011 for details). The energy released by the accretion shocks on the star, as well as the high-energy particles and radiation produced in the reconnection layers control the evolution of the disk through photoevaporation and provide energy for the disk chemistry.

as reconnection events at the base of the jet. Disk winds result from the heating by the stellar XUV radiation of the gas in the atmosphere of the protoplanetary disk. A thermal flow is initiated at locations where the sound velocity overcomes the escape velocity. In addition, if the disk has a significant magnetic field, the thermal flow is aided by the poloidal (out of the disk) component of the field to overcome gravity. An MHD flow is generated in this manner which is centrifugally driven (as shown in Fig. 2.10). The most successful attempts to reach the base of the jet have been carried at UV wavelengths for two reasons: the thermal regime (from few thousand to 30,000 K) and the high angular resolution.

The strong UV lines of C IV, Si III], C III], C II] and Mg II have been used to probe the warm base of the jet and to study the jet collimation mechanisms (Gómez de Castro and Verdugo 2001, 2003; Skinner et al., 2018). Numerical simulations predict that significant Si III], C III] radiation should also come from the current layer between the star and the disk (see Fig. 2.10) however, comparison between the observed profiles and the theoretical predictions suggests that the unresolved emission around the star is dominated by the stellar magnetosphere and the accretion shocks, see Fig. 2.11 (Gómez de Castro and von Rekowski 2011). Unfortunately, UV observations of the optically thin Si III] and C III] lines are too scarce to enable a comprehensive study. Moreover, theoretical modelling needs to be refined; current numerical models are based in 2.5-D simulations, i.e., cylindrical symmetry is imposed to the problem and magnetic reconnection is not properly dealt with (see i.e., Ripperda et al., 2017).

On the larger scales, UV instrumentation provides the best possible resolution to study jet rotation (Coffey et al., 2015). Moreover, the interaction of the jet with the surrounding environment leads to the formation of shock excited nebulosities that are observed at UV wavelengths provided the source is not deeply embedded in the cloud (see Gómez de Castro and Robles 1999 for a compilation of UV observations of Herbig-Haro objects and jets). Strong radiative shocks are produced at the bow shocks where most of the kinetic energy is released into heating; among the main results obtained from UV observations is the realization that magnetic fields ought to be softening the collision at the working surface of the jet since high excitation lines, such as OVI, are not observed (Raymond et al., 1997). This effect could also be caused by the high clumpiness of Herbig-Haro objects consisting of dense hot clumps (T = 10^5 K, $n_e = 10^6$ cm^{-3}) embedded within a warm flow (T = 10^4 K, $n_e = 10^3$ cm^{-3}). The filling factor of the hottest component is ~0.1−1%; it was derived from UV-optical monitoring observations (Liseau et al., 1996).

3.1.2 UV radiation from the disk

The lifetime, spatial distribution, and composition of gas and dust of young (age < 30 Myr) circumstellar disks are important properties for understanding the formation and evolution of extrasolar planetary systems. Disk gas regulates planetary migration (Ward, 1997; Armitage et al., 2002, Armitage, 2007; Trilling et al.,

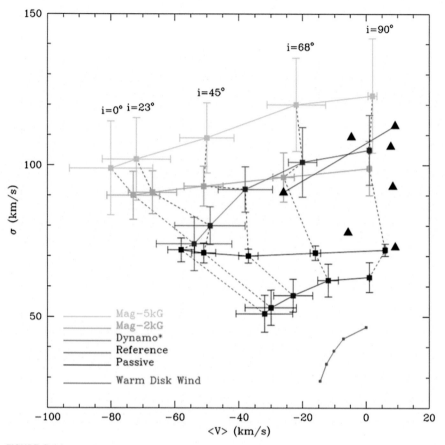

FIGURE 2.11

Diagram showing the expected width and centroid of the profiles of the Si III] UV line formed in the TTSs outflows for various possible ejection mechanisms: star-disk interaction and warm ($T_e \sim 10^4$ K) disk winds. Outflows driven by the star-disk interaction have two regimes: quiescent and eruptive; during the eruptive phase, mass is ejected from the current layer after reconnection. The values represented in the plot correspond to the quiescent phase. The time average values of the profile broadening (σ) and centroid (<V>) are represented as squares, while the error bars indicate true temporal variations predicted by theory during quiescence. Purple squares at the bottom right represent predictions from simple warm disk wind models (Gómez de Castro and Ferro-Fontán, 2005). Observed values obtained from the few measurements of TTSs (DE Tau, AK Sco, RY Tau, RW Aur, T Tau and RU Lup) are plotted as big black triangles. The observed line broadening agrees well with the predictions from magnetospheric ejection models but no strong dependence on inclination is observed. Objects like RU Lup and RY Tau with inclinations 24° and 86°, respectively, have similar dispersions.

After Gómez de Castro and von Rekowski (2011).

2002) and the migration timescale is sensitive to the specifics of the disk surface density distribution and dissipation timescale. Moreover, the formation of giant planet cores and their accretion of gaseous envelopes occurs on timescales similar to the lifetimes of the disks.

Typical lifetimes of proto planetary disks are about 2−3 Myr (Williams and Cieza, 2011) hence, in a 5 Myr old young cluster around 90% of the disks should be dispersed (Sicilia-Aguilar et al., 2006; Hernández et al., 2007). High angular resolution observations of protoplanetary disks obtained with ALMA show consistently gaps in the dust distribution at these early ages; these gaps are thought to be carved by nascent planets (ALMA partnership, 2015). The accretion flow through these voids is expected to be low, if existent at all, and this is difficult to reconcile with the simultaneous observation of powerful bipolar outflows in the same sources (see i.e., Dipierro et al., 2015). The prototype of this problem is shown in Fig. 2.12. HL Tau's disk has been mapped with the highest possible angular resolution (0.03 arcsec or 5 AU) and sensitivity; several rings or gaps are observed in the dust distribution at projected radii ranging from of ~ 20 AU to 190 AU. The accretion rate onto HL Tau, a 1 Myr old TTS, is measured to be $\sim 7 \times 10^{-7} M_o$ yr^{-1} (Lin et al., 1994) and powers a large-scale jet that excites several Herbig-Haro nebulosities.

Another puzzling issue relates with the low CO (gas) emission detected on those scales by ALMA which is much weaker than expected from the dust emission and the few detections of HD (Bergin et al., 2013; McClure et al., 2016). The most plausible explanation is that complex C-bearing molecules freeze-out in the disk midplane. Once frozen out in a low-viscosity region, the ices will remain frozen out, with a time-dependence that will gradually deplete the disk of C and therefore CO (e.g., Kama et al., 2016; Schwarz et al., 2016; Yu et al., 2016).

Observations of the inner few AUs of protoplanetary disks are crucial to address these problems. Jets and disk winds are ejected from the innermost AUs thus, large scale disk processes will not affect them immediately. Surveys of CO, H_2O, and organic molecules from the inner few AU have provided new constraints on the radial distribution, temperature, and composition of planet-forming disks (e.g., Salyk et al., 2008, 2011; Carr and Najita 2011; Brown et al., 2013; Banzatti et al., 2015) however, the UV emission of H_2 is the most sensitive probe since reaches gas column densities $<10^{-6}$ g cm^{-2} thus making feasible the detection of tenuous gas in the protoplanetary environment. While mid-IR CO spectra or other traditional accretion diagnostics suggest that the inner gas disk has dissipated, far-UV H_2 and CO observations offer unambiguous evidence for the presence of an inner (r < 10 AU) molecular disk that persists to 10 Myr (France et al., 2011, 2012b).

Moreover, the CO UV bands together with those of H_2 can be used to measure the CO/H_2 ratio in the inner disk and confront them with the larger scale CO data (30−100 AU) being provided by ALMA.

UV spectroscopic observations are also fundamental to detect the volatiles released by dust, planetesimals and comets at later stages of the evolution, in young

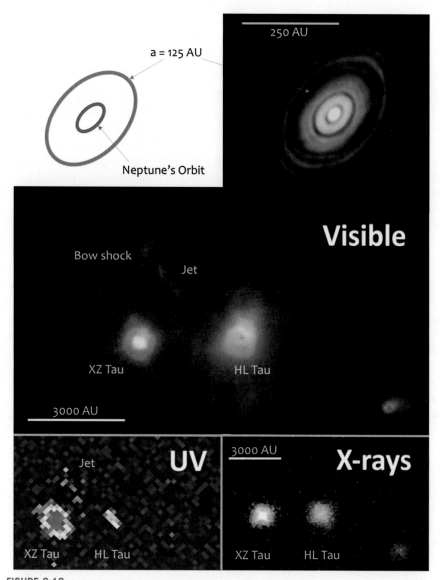

FIGURE 2.12

Observations of the 1 Myr old TTS HL Tau (spectral type K5). In the central panel, the visible observations (SDSS9) show the star, its neighbor (XZ Tau), the collimated jet and hints of the bow shock which is better observed in the main emission lines. The jet and the stars are observed at UV wavelengths (GALEX) and the stars and some hints of the jet in X-rays (Chandra). In the upper right inset, the reconstructed image of the disk from interferometric data obtained with ALMA is displayed. This is the best angular resolution image ever obtained of a young proto planetary disk; 0.035 arcsec or 5 AU. The innermost region of the disk where Earth like planets form (<1AU) can only be observed by spectroscopic means through the UV emission of H_2.

planetary and debris disks. The abundances of the vaporizing species and the main photochemical and physical processes acting on the inner part of the disk have been measured in this manner (Vidal-Madjar et al., 1998; Lecavelier des Etangs et al., 2001, 2003; Roberge et al., 2014; Miles et al., 2016).

3.1.3 Stellar UV radiation from the TTS phase to the main sequence

UV radiation from late type stars is the main tracer of their magnetic activity; magnetic energy is transported from the stellar surface onto the atmosphere heating it. The energy is dissipated into three main regions: the chromosphere ($T \sim 10^4$ K), the hot corona reaching temperatures of MK and the transition region (TR) between them. The TR is a very thin layer only observable at UV wavelengths. There are well characterized correlations between the flux radiated in the various spectral tracers of these regions that are used to model energy transport in cool stars and call for a universal mechanism operating in them (Ayres et al., 1995; Mihalas, 1978).

In TTSs, these signatures are overcome by the UV radiation released in accretion shocks that has a markedly different SED. Accretion shocks are produced at the points where the infalling material collisions with the stellar surface. Matter at the inner border of the disk gets trapped by the stellar magnetic field and channeled onto open field locations on the stellar surface (Romanova et al., 2004, 2012); free-fall speeds are around $300-400$ km s^{-1} driving at temperatures of ~ 1 MK at the shock front. The reprocessing of the X radiation produced at the shock front to lower energies in the subsequent recombination cascade is the source of the observed UV radiation from accretion shocks (Gómez de Castro and Lamzin, 1999; Gullbring et al., 2000). The difference between the flux-flux relation in late-type main sequence and TTSs is displayed in Fig. 2.13. The main difference is the excess of low ionization species that it is caused by the high abundance of warm matter in the immediate vicinity of the star in the TTSs.

UV radiation during the T Tauri phase is typically 50 times stronger than during the main sequence evolution (Johns-Krull et al., 2000; Yang et al., 2012; Gómez de Castro and Marcos-Arenal, 2012; López-Martínez and Gómez de Castro, 2015). Surface normalized fluxes of some relevant spectral lines are displayed in Fig. 2.14. Main sequence cool stars (G-K) have surface normalized fluxes typically between 10^{-7} and 10^{-6} that raise by one to two orders of magnitude in the very active M-type stars. TTSs normalized fluxes are higher by another one to two orders of magnitude, the lowest values are observed in the more evolved, nonaccreting TTSs and are comparable to those of M-type stars.

The profile of the spectral lines is very broad however, the broadening does correlate neither with the stellar rotation velocity nor with the inclination of the TTS (López-Martínez and Gómez de Castro, 2014); in the PMS star RW Aur, the broadening seems to be caused by an ion torus (Gómez de Castro and Verdugo, 2003). Line broadening is observed consistently in all species, ranging from 10^4 K to 10^5 K plasma tracers, and decreases as the star approaches the main sequence (Ardila et al., 2013; Gómez de Castro, 2013b; López-Martínez and Gómez de Castro, 2014).

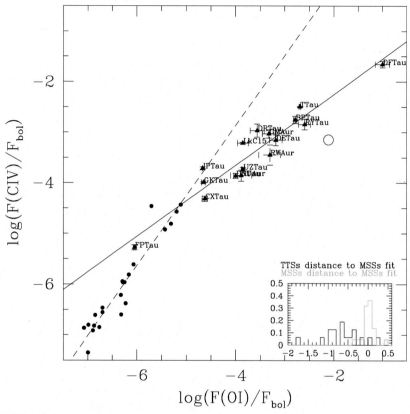

FIGURE 2.13

The C IV flux is plotted versus the O I flux, both normalized to the stellar bolometric flux, for the TTSs (named triangles) and compared with main sequence stars (black dots). The large circle marks the location of the brown dwarf 2MASS J12073346-3332539 (France et al., 2010). The flux-flux regression lines are plotted for TTSs and main sequence stars with a solid and a dashed line, respectively. The bottom right inset displays the distribution of the distances of the stars in both samples to the regression line defined by the late-type main sequence stars.

After Gómez de Castro and Marcos-Arenal (2012).

The profiles of warm ($T \sim 10^3 - 10^4$ K) plasma tracers such as the Mg II lines, show, in addition, the absorption produced by a warm outflow. When the terminal velocity of the wind is measured, it is found it to be always smaller than the escape velocity from the stellar surface suggesting that the absorption is produced by the disk wind (López-Martínez and Gómez de Castro, 2014).

As pointed out in Section 2, stellar FUV emission is a critical input for chemical models of protoplanetary disks and to estimate photoevaporation (Gorti et al., 2009; Owen et al., 2010). Lyα emission accounts for 80% of the FUV emission

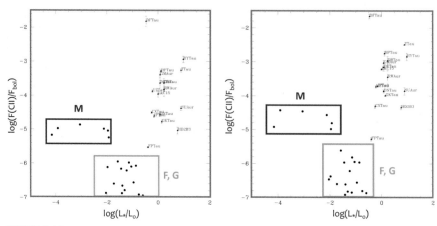

FIGURE 2.14

CII and CIV surface normalized fluxes versus the stellar luminosities for some TTSs compared with main sequence late type stars (circles; red square for M type stars; blue square for F and G type stars). Error bars for the UV normalized fluxes of the TTSs are indicated.

(Schindhelm et al., 2012) and may penetrate deeper into the disk by resonant scattering (Bethell and Bergin, 2011), thereby dissociating species such as H_2O, HCN, and other species (e.g., Bergin et al., 2003; Fogel et al., 2011). Moreover, the UV radiation from the star contributes to ionization of the disk which is relevant for the onset of nonideal magnetic effects in the disk and accretion (Bai, 2017).

Unfortunately, the UV observations of low mass, low accretion rate TTSs are scarce. For technical reasons, Earth-like exoplanets are being searched at the low mass end of the main sequence and it is fundamental to determine the evolution of the radiation environment during their PMS evolution, at the time when the planet and its atmosphere are being formed.

4. Evolution of planetary atmospheres

The capability of a planet to support stable liquid water on its surface, and thus meet the requirements for habitability (Kasting et al., 1993; Lammer et al., 2009; Kopparapu et al., 2013), is conditioned by the presence of an atmosphere that provides the proper conditions, such as pressure, temperature, shield against the stellar radiation, and greenhouse warming, protecting the water inventory of the planet.

Formation and evolution of Earth-like atmospheres involves long-time processes, from H and He accumulation from the stellar nebula in the very early stages of planetary assembly, resulting in H-rich protoatmospheres, to the formation of secondary atmospheres due to ocean magma solidification and strong outgassing due to high tectonic activity (Lammer, 2013). Whether a planet could sustain its atmosphere depends not only on the initial inventory of gas (Lammer et al., 2014), but also on the action of the incident stellar UV radiation and plasma winds coming from the parent star.

UV radiation of the host star controls the main processes taking place in the upper atmospheric layers of a planet. As confirmed by the first observations of the atmosphere of hot Jupiters, the opacity of the atmosphere decreases with the wavelength of the incident radiation (e.g., Vidal-Madjar et al., 2003; Lecavelier des Etangs et al., 2010; Poppenhaeger et al., 2013). Thus, EUV radiation is the responsible of the reactions of the uppermost layers of the atmosphere, while FUV and NUV (250–320 nm) radiation can penetrate deeper in the atmosphere (see Fig. 2.15).

The X-ray and UV (XUV) radiation directly interacts with the atoms populating the uppermost layers of planetary atmospheres. These interactions include atomic hydrogen photoionization, heating, atmospheric expansion, and particle losses through thermal and nonthermal escape.

Escape processes are especially important at the early evolutionary stages of planets orbiting solar-like stars. At that time, XUV radiation of the host star is suggested to be much more intense than the present solar values. Ribas et al. (2005) used the X-ray and UV spectra of 6 G-type stars included in the *Sun in Time* program, with ages ranging from 0.1 to 6.7 Gyr, to show that young, main-sequence solar like stars exhibit XUV emissions ∼ 100–1000 times stronger than those of the present Sun. Similarly, the chromospheric FUV emission of these young solar-like stars

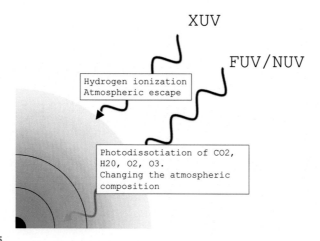

FIGURE 2.15

Action of XUV and FUV/NUV radiation at different levels of the planetary atmosphere.

is expected to be 20–60 times stronger than at present (see Fig. 2.16). Moreover, young solar-like stars emit intense flares and superflares that enhance considerably the EUV flux received by the planet (see also Estrela & Valio contribution in this volume).

Planets orbiting cooler M dwarf stars are heated by significantly stronger XUV fluxes (Poppenhaeger et al., 2010). These objects are the most numerous stars in our galaxy, and most Earth-like planets have been detected orbiting around them (Dressing and Charbonneau, 2015), including the well-studied Proxima b (Anglada-Escudé et al., 2016) and TRAPPIST-1 planetary systems (Gillon et al., 2017). The total flux coming from these faint objects is orders of magnitude lower than from solar-like stars; however, their SED is quite different from that of the Sun. For instance, the EUV flux reaching Proxima b, in orbit around Proxima Centauri, is about ~ 30 times higher than the present day solar value at one AU (Ribas et al., 2017), as displayed in Fig. 2.17. Due to the slow evolution of M type compared to solar-like stars, this enhanced XUV phase can be extent over several billion years (West et al., 2008; Ramirez and Kaltenegger, 2014).

XUV radiation, with photon energies from 10 eV up to 124 eV, ionizes the atmospheric hydrogen, creating a population of photoelectrons that collisionally heat the atmospheric atoms, increasing their thermal velocity. In the exosphere of the planet, defined as the region where the density is too low for collisional processes to be relevant, particles having thermal velocities greater than the escape velocity of the planet are susceptible to escape.

Depending on the ratio between the gravitational potential at the atmosphere and the thermal energies reached by the atmospheric particles different mass loss regimes emerge ranging from the moderate, classical Jeans escape, where particles

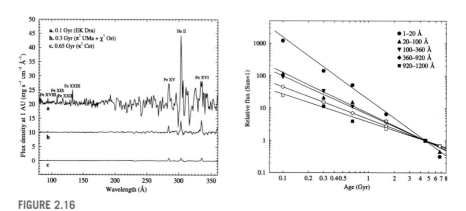

FIGURE 2.16

Left: EUV irradiances for stars at different evolutionary stages, showing a notably decrease in the strength of the emission lines. Right: Relative flux to present solar values for different stellar ages and wavelength bands.

From Ribas et al. (2005).

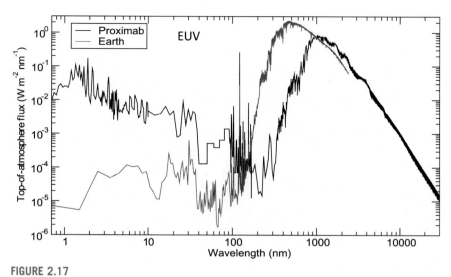

FIGURE 2.17

Fluxes at the top of the atmospheres of the Earth (in red, at 1 AU) and Proxima b (in black, at 0.0485 AU).

From Ribas et al. (2017).

in the high-energy tail of the Maxwell distribution are lost, to the stronger hydrodynamic blow off regime. In this last scenario, the thermosphere is heated to the point that the hydrostatic condition of the atmosphere is broken leading to hydrodynamic expansion of the atmosphere up to several planetary radii (Tian et al., 2005, 2008; Lammer et al., 2008, 2013; Linsky, 2019). This process may trigger the rapid evaporation of the entire atmosphere; between both mechanisms, an intermediate escaping processes, the so-called boil-off or strong Jeans escape could also drive the particle losses from the planetary atmosphere. The thermal escape mechanism will be conditioned then by the gravity of the planet, the molecular mass of atmospheric species, the incident EUV radiation and the efficiency of the EUV flux to heat conversion. Thermal escape driven by the XUV radiation of the host star has dramatic consequences in the formation and evolution of Earth-like atmospheres, conditioning the survival of these atmospheres and the emergence of life.

Erkaev et al. (2013) studied hydrodynamic escape in H-rich atmospheres of Earth-like and super-Earth-like planets (size of 2 Earth radius and a mass of 10 Earth masses) under different levels of XUV radiation, from present solar levels up to 100 times that value. Under the most intense XUV flux, planets may expand their exobase levels up to 20 planetary radii. Moreover, they showed that an Earth-like planet under extreme XUV flux conditions may lose completely its primordial H-rich atmosphere, while more massive super-Earth planets could retain some fraction of its H atmosphere. This is in agreement with other studies, in which hydrodynamic models predict an atmospheric loss in time scales about 100 Myr for Earth and sub-Earth like planets under high XUV stellar radiation (Lammer et al., 2014;

Kubyshkina et al., 2018) and high stellar rotation activity (Johnstone et al., 2015), or planetary distance (Erkaev et al., 2016).

As EUV/FUV radiation dominates the dynamics of the particles populating the upper layers of a planetary atmosphere, driving atmospheric escape, NUV radiation penetrates deeper into the planetary atmosphere, interacting with atoms, ions and molecules in the lower atmospheric layers. These interactions include photodissociation and ionization reactions, that may change the composition of the planetary atmosphere, including the planetary O_2 and O_3 shield, protecting the biological activity on the planetary surface.

O_2 and its photochemical byproduct O_3, are considered the most important exoplanetary biosignatures due to its biological origin on Earth (Des Marais et al., 2002; Kaltenegger et al., 2007); both exhibit large photodissociation cross-sections to FUV-NUV radiation (e.g., Hu et al., 2012 and references therein). This is also the case of other crucial molecules such CO_2 and H_2O, that are also suggested to be major constituents of Earth-like exoplanetary atmospheres (see Fig. 2.18).

According to their wavelength-dependent photodissociation cross-sections, FUV photons photodissociate CO_2 and O_2 to form atomic oxygen via the reactions:

$$CO_2 + h\upsilon(\lambda < 175 \text{ nm}) \rightarrow CO + O$$

$$O_2 + h\upsilon(\lambda < 240 \text{ nm}) \rightarrow O + O$$

Atomic oxygen then may recombine to form O_2, and may also recombine with molecular oxygen to form ozone in three-body reactions (including a third molecule, M, needed to carry off the excess energy), leading to ozone formation:

$$O + O_2 + M \rightarrow O_3 + M$$

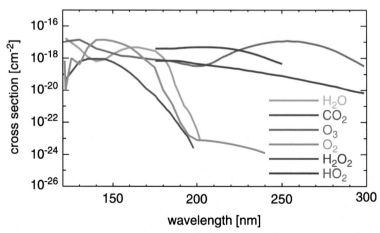

FIGURE 2.18

Photodissociation cross-sections for important Earth-like atmospheric molecules in the FUV-NUV spectral range.

From Meadows (2017).

Thus, the FUV radiation from the host star may produce significant amounts of photochemical (no biotic) O_2 and O_3, in CO_2 and O_2-rich atmospheres, and also in water-dominated atmospheres (Wordsworth and Pierrehumbert, 2014; Luger and Barnes, 2015) from H_2O photolysis and subsequent H loss.

Also, NUV radiation will photodissociate O_3 molecules, following the reaction:

$$O_3 + h\upsilon(\lambda < 340 \text{ nm}) \rightarrow O_2 + O$$

In this sense, the abiotic production of O_2 and O_3 seems to be controlled by the FUV and NUV radiation from the parent star. Thus, an accurate characterization of the FUV and NUV emission of the star is indispensable to identify the production of abiotic oxygen and ozone, resulting in false-positive detections of these biosignatures.

4.1 UV observation of exoplanets

4.1.1 Earth as an exoplanet

To date, the Earth constitutes our only reference as a living planet. An accurate characterization of the main spectral features of the terrestrial atmosphere is crucial for (1) identifying and interpreting the Earth's life fingerprints and (2) constraining the atmospheric models that will predict the spectrum of the atmospheres of other planets beyond the solar system, within a wide range of planetary (mass, radius, bulk composition) and stellar characteristics (incident stellar radiation, stellar age, luminosity, spectral type, and so on).

Some of the most relevant biosignatures and habitability tracers of the Earth's spectrum reside in the UV part of the electromagnetic spectrum, including ozone, sulfur molecules and Rayleigh and cloud scattering. However, throughout the 4.5 Gyr history of life on Earth, the shape and main biosignatures present in the UV spectrum of our planet have substantially changed, as life on Earth coevolved with the geological processes. The spectrum of an ancient Earth could give hints of how a planet with no apparent clear signs of life undergoes evolutionary processes to develop life signs on its surface or in its atmosphere. In this sense, the UV range of the spectrum constitutes a powerful tool to investigate the evolutionary bio-state of an Earth-like planet.

In this section, we describe the main UV tracers of our planet's atmosphere, observations of the Earth, modeling and evolution of the main biosignatures in the UV spectrum.

4.1.1.1 Earth's observations in the UV range

Earth spectra obtained from interplanetary spacecrafts provide the best representation of the Earth as an exoplanet. From the early UV image of the Earth taken from the Moon by the Apollo XVI mission with Carruthers' telescope (Carruthers, 1973), several missions have registered data of the Earth atmosphere at UV wavelengths, including the Magnetopause-to-Aurora Global Exploration (IMAGE) satellite (Fuselier et al., 2000), the Two Wide-Angle Imaging Neutral-Atom Spectrometers (TWINS, McComas et al., 2009), the EPOXI Discovery Mission of Opportunity using the Deep Impact flyby spacecraft (Livengood et al., 2011; Robinson et al., 2011; Schwieterman et al., 2015), the *Lunar CRater Observation and Sensing Satellite* (*LCROSS*, Robinson et al., 2014), Rosetta/VIRTIS (Visible and Infrared Thermal Imaging Spectrometer, Hurley et al., 2014), Mars Global Surveyor/THS (Thermal Emission Spectrometer, Christensen and Pearl, 1997) or Deep Space Climate Observatory (Yang et al., 2018).

Due to its spectral coverage (0.26−13.5 μm), the LCROSS mission provided data of the best UV tracers of the Earth's biotic activity: O_3. LCROSS Earth's spectrum shows the imprint of ozone, absorbing in the Hartley-Huggins bands (between 150 and 350 nm, see Fig. 2.19). These bands are very sensitive to the column density of O_3 and saturate with an abundance of ∼ 1 ppm or less (see Fig. 2.20). In the spectrum of the Earth, this results in a saturated spectrum at wavelengths shorter than 320 nm since the concentration of O_3 is ∼ 10 ppm in the stratosphere between 15 and 30 km in altitude.

Beside ozone, the UV continuum produced by Rayleigh scattering caused by the N_2, H_2O and CH_4 molecules, is an important bio-tracer. The slope of the Rayleigh scattering shows a clear dependence on the surface pressure of the planet, a key factor in terms of the possible habitability of a planet (Feng et al., 2018).

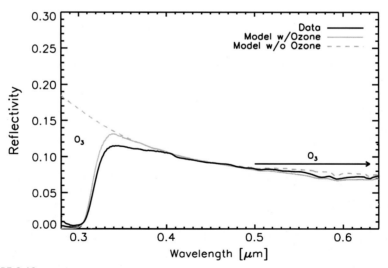

FIGURE 2.19

Reflectivity of the Earth in the [0.3−0.6 μm] band, showing data from LCROSS observations (solid black), model with ozone (solid gray) and model without ozone (dashed gray).

Adapted from Robinson et al. (2014).

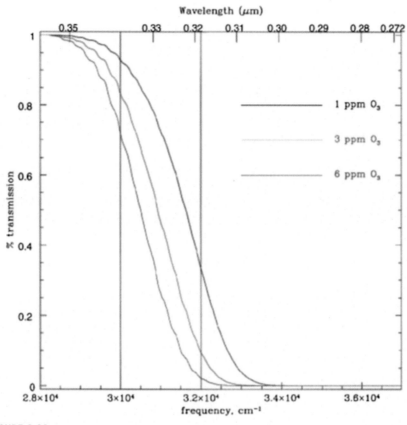

FIGURE 2.20

Reflection UV spectrum model of the Earth for three abundance levels in the stratospheric ozone layer.

After Des Marais et al. (2002).

Even though the presence of atmospheric sulfur aerosols is linked to the geologic activity rather than a biological origin, the presence of these gases is also crucial in terms of habitability, as they are tracers of an active vulcanism (Hu et al., 2013; Kaltenegger and Sasselov, 2010; Misra et al., 2015). H_2S and SO_2 produce spectral absorptions in the UV range at wavelengths $<0.3\,\mu m$ (Robinson et al., 2014), as well as its photochemical byproduct the cyclic octatomic molecule S_8. However, these spectral features are not detectable in the present spectrum of the Earth, due to their short chemical timescales (Seinfeld and Pandis, 2016; Hu et al., 2013).

4.1.1.2 The Lyα footprint of the Earth

The Earth's exosphere—the low-density, collisionless, outermost region of the atmosphere—is mainly populated by neutral hydrogen atoms. Depending on their trajectories, exospheric particles could be described as escaping particles (with velocities higher than the escape velocity of the planet, lost from the planetary atmosphere), ballistic (reaching a maximum altitude and falling down to the exobase), and satellite particles (ballistic particles undergoing collisions, so they do not fall into the exobase, Chamberlain, 1963; Beth et al., 2016).

The best method to detect the presence of these particles in the high atmosphere of the Earth is trough UV observations, as the Lyα line is the most sensitive tracer of the presence of atomic hydrogen. As H I atoms absorb the Lyα photons coming from the Sun, the solar radiation is blocked during the transit of the planet in front of the star.

Despite the difficulty of mapping this region through Lyα measurements from distances far from the Earth's surface, and the problematic presence of a secondary Lyα emission from H atoms in the interplanetary medium, several missions have compiled data to describe the extension and properties of this rarified envelope. The first observations of the terrestrial exosphere carried out by the Mariner five missions at a distance of 37 R_E (Wallace et al., 1970) and by Apollo 16 (Carruthers and Page, 1972) revealed an exospheric extension of ∼15 planetary radii. More recent observations carried out by the GEO instrument on board the IMAGE satellite (Fuselier et al., 2000) found that the hydrogen density distribution is essentially cylindrically symmetric around the Sun-Earth line, exhibiting an enhancement in the antisolar direction, toward the geotail (Mende et al., 2000). The density distribution of the exospheric particle population was found to be bimodal, with a dominant central component and an extended component peaking at ∼8R_E. Later on, the TWINS mission (McComas et al., 2009) reported enhancements by a factor of ∼2 to 3 of the extended component with respect to GEO measurements. Observations from the SWAN-SOHO mission (Baliukin et al., 2019) and the Lyman Alpha Imaging Camera on board the Japanese spacecraft PROCYON (Kameda et al., 2017), further confirmed the presence of this extended component, revealing an exospheric extension of more than 38 Earth radii (see Fig. 2.21).

The extension of the Earth's exosphere has been analytically modeled, considering that the dynamics of the H I exospheric atoms is controlled by the gravity influence of the planet and the radiation pressure of the Sun (Chamberlain, 1979; Bishop and Chamberlain, 1989; Beth et al., 2016).

In a secondary atmosphere such as that of the Earth, exospheric atomic hydrogen is produced by photodissociation reactions of water, methane, and/or molecular hydrogen at lower atmospheric layers, in turn diffused into the upper atmosphere (Dessler et al., 1994, Brasseur and Solomon, 1996). In this sense, the detection of a hydrogen envelope around a distant planet could be used as an indirect measurement of the presence of greenhouse effect gases, indispensable for the planetary habitability on Earth-like exoplanets. As suggested in Selsis et al. (2007) and Bolmont et al. (2017), ocean worlds may also have extended H-rich exospheres fed by water evaporation.

The observation of the large extent of the Earth's exosphere showed the limitations of current theoretical models to foresee the expected properties of exo-Earths

FIGURE 2.21

Distribution of the exospheric neutral hydrogen captured by the PROCYON spacecraft, showed in Rayleigh on a logarithmic scale.

After Kameda et al. (2017).

and hence, their detectability (see Fig. 2.22). With these provisions, Gómez de Castro et al. (2018) showed the feasibility of detecting Earth-like exoplanets around M-dwarf stars with moderate size (four to eight m primary mirror) space observatories through transit spectroscopy in the Lyα line. This result was later confirmed by Kislyakova et al. (2019) and Dos Santos et al. (2019).

4.2 Earth's UV biosignatures on time

As shown from Earth observations and models, the ozone absorption in the Huggins-Hartley band is the most remarkable biosignature in the UV spectrum of our planet. Ozone is produced by the photodissociation of oxygen via the so-called Chapman reactions (Chapman, 1930, see Section 2), and its abundance can be scaled to that of the oxygen (see Fig. 2.23, Kasting and Donahue, 1980; Reinhard et al., 2017).

The presence of ozone in the Earth's atmosphere is then susceptible to changes in the amount of oxygen present in the atmosphere and the UV photoionizing radiation coming from the star. As shown in Fig. 2.24, the levels of oxygen on Earth have undergone dramatic changes over the history of Earth (Lyons et al., 2014; Planavsky et al., 2014; Olson et al., 2018). In the timespan between 4.0 and 2.5 Ga ago, corresponding to the Archean era, levels of oxygen are suggested to reach only 10 millionths of the present atmospheric level (PAL), Zahnle et al., 2006), or even lower

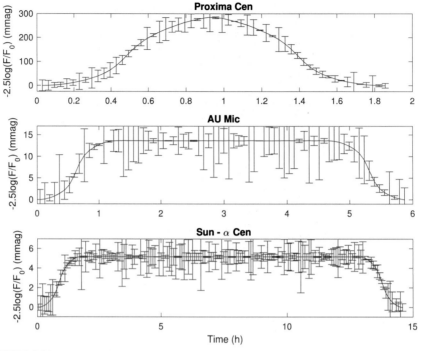

FIGURE 2.22

Predicted Lyα light curves for an Earth transiting a star like the Sun, and M0 V star and Proxima Cen (M5.5 V); no ISM absorption is considered (see Gómez de Castro et al., 2018 for the details of the calculations). The calculations show that SNR(Proxima Centauri) = 11.7 can be reached with a 4 m telescope and time binning of 2 min. However, to reach SNR = 3.5 observing an M0 star at AU Mic distance, it is necessary to monitor it with a 12 m telescope with a time binning 4 min. To observe the Earth's transit in front of the Sun from a star at the distance of α-Cen with SNR = 4.4, a 12 m telescope and a time binning 10 min are needed.

After Gómez de Castro et al. (2018).

values of $\sim 10^{-7}$ PAL according to the models in Claire et al. (2006) and Goldblatt et al. (2009). The Great Oxidation Event (GOE, Holland, 2002; Luo et al., 2016) at the beginning of the Proterozoic era (2.5−0.5 Ga ago) notably increased the amount of oxygen in the terrestrial atmosphere, leading to oxygen concentrations values from 0.1 PAL (Planavsky et al., 2014, 2018), up to 10%−40% of the current concentration (Kump, 2008) during the mid-Proterozoic.

Only for a small fraction of the Earth's lifetime, from 0.5 Ga ago on, the levels of oxygen reached the levels of the current terrestrial atmosphere, corresponding to the develop of animal life on the Earth's surface (Reinhard et al., 2016).

The strong O_2 Fraunhofer A band, observed at optical wavelengths (760 nm), is a sensitive probe to detect atmospheric oxygen during the last 10th of the Earth's history. However, the great sensitivity of the Hurtley-Huggins band of ozone in the UV spectrum of the Earth (saturated for values greater than one ppmv) makes ozone detectable in the most part of the Earth's evolutionary lifetime. This fact has a great

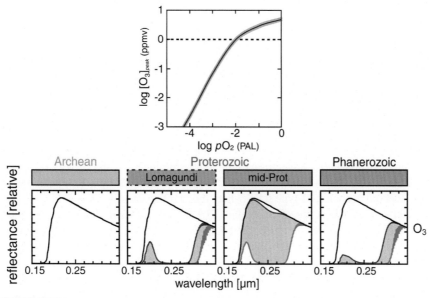

FIGURE 2.23

Top: Values of peak atmospheric O_3 as a function of ground-level pO_2 according to the results of Kasting and Donahue (1980). Bottom: Reflectance spectrum of the O_3 Huggins-Hartley band (0.25 μm) at different geologic times. Lower and upper abundance limits are given in red and blue respectively, and the region between these limits is shaded gray. The black line represents the case with no O_3 absorption.

Adapted from Reinhard et al. (2017).

impact in the possible false-negatives detections of oxygen, making ozone a powerful biosignature on planets with very-low, no-detectable levels of oxygen present in the atmosphere.

In contrast to the O_2 and O_3 abundances, the abundance of methane in Earth's atmosphere during the most of Hadean and Archean eons was suggested to reach

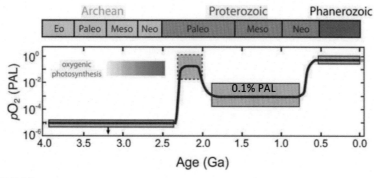

FIGURE 2.24

Evolution of oxygen concentration on Earth in terms of the present atmospheric level (PAL) of O2.

After Schwieterman et al. (2018).

levels 1000 times those of the present atmospheric levels (Catling et al., 2001; Claire et al., 2006; Zahnle et al., 2006; Haqq-Misra et al., 2008). These enhanced levels may trigger the formation of hydrocarbon hazes (Domagal-Goldman et al., 2008; Arney et al., 2016). These hazes absorb very efficiently the UV radiation, and could be also used as a potential biosignature (see Fig. 2.25, Arney et al., 2018).

4.2.1 Exoplanets

4.2.1.1 Exoplanetary upper atmospheres

Observation of the upper atmosphere of exoplanets is accessible through FUV transit spectroscopy, showing the light of the host star that is transmitted through the planetary atmosphere. The UV spectrum of the Sun is dominated by the strong Lyα emission line that carries on as much energy as the rest of the FUV flux. This behavior is also observed in cooler, M-type stars, where the Lyα line constitutes $\sim 37\%-75\%$ of the total 1150–3100 A flux (France et al., 2012a, 2013). Due to this large contribution to the UV flux, the variation of the Lyα profile during transits constitutes a powerful tool to determine the presence of extended hydrogen envelopes around exoplanets.

Evidence from the presence of extended H-rich planetary exospheres have been reported for giant, gaseous exoplanets orbiting at close distances of their parent star. This is the case of the Hot Jupiters HD 209,458b (Vidal-Madjar et al., 2003; Ehrenreich et al., 2008; Linsky et al., 2010) and HD 189,733 b (Lecavelier Des Etangs et al., 2010, 2012), and the warm Neptunes GJ 436b (Kulow et al., 2014; Ehrenreich et al., 2015) and GJ 4370b (Bourrier et al., 2018). The enhanced absorption in the wings of the Lyα line during the planetary transit revealed the presence of a strong hydrogen escape from the atmosphere of these planets.

Besides the detection of extended hydrogen atmospheres, observations in the UV wavelength range also offers the possibility to detect heavier compounds. At high XUV radiation levels, hydrodynamic escape can be so strong that could drag the heavier atoms present in the atmosphere. This is the case of HD209458b, showing C, O, and Si absorption (Vidal-Madjar et al., 2004; Linsky et al., 2010) in its FUV spectrum, and Mg at longer wavelengths in the NUV spectrum (Vidal-Madjar et al., 2013). Heavy metals such as Mg and Fe also have been detected in

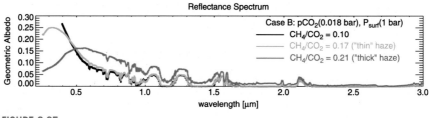

FIGURE 2.25

Spectrum of Archean Earth with three different haze thicknesses (Arney et al., 2016).

the Hot Jupiter WASP-12b (Fossati et al., 2010; Haswell et al., 2012) and Fe II in WASP-121b (Salz et al., 2019).

However atmospheric escape has not been detected to date from Earth-like exoplanets at UV wavelengths, such as Cnc55 (Ehrenreich et al., 2012), HD97658b (Bourrier et al., 2017a) or the Kepler 444 planetary system (Bourrier et al., 2017b). Masses and radius measurements of Kepler planets have revealed planetary densities suggesting that these planets could sustain atmospheres with different compositions to that of the solar system rocky planets. A large fraction of them have densities compatible with H/He rich envelopes surrounding rocky cores (Wu and Lithwick, 2013; Hadden and Lithwick, 2014; Rogers and Seager, 2010). This is the case of the well-studied planetary system Kepler-11 (Lissauer et al., 2011) or GJ 1214b (Charbonneau et al., 2009), which seem to retain remnants of their primordial hydrogen-dominated atmospheres.

Whether a planet could retain part of its primordial H atmosphere against the incident stellar flux has crucial consequences for detectability. Kepler exoplanets census has revealed that the presence of $2-4\,R_\oplus$ planets is dramatically reduced for orbits <7 days (e.g., Howard et al., 2012; Sanchis-Ojeda et al., 2014; Lundkvist et al., 2016), suggesting that XUV-driven escape-processes have been crucial to remove the gaseous H, He envelopes of these planets, leading to naked rocky cores (see Fig. 2.26; Lopez et al., 2012; Lopez, 2017; Owen and Wu, 2013, 2017; Jin et al., 2014).

Beside the great advantage to use the Lyα transmission spectroscopy to detect the presence of H-rich extended envelopes around exoplanets, the measurement of the transmission Lyα spectrum also provides crucial information about the planet, such as the escape rate, the composition, density, velocity of the escaping particles,

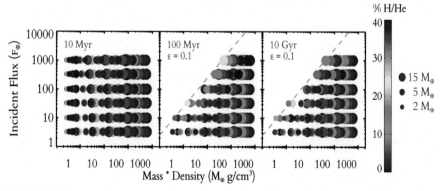

FIGURE 2.26

Prediction of the two to four Earth radius absence in Kepler planets for close-in orbits in Lopez et al. (2012). Planets with masses from 2 to 15 Earth masses with different atmospheric H/He masses are exposed to stellar fluxes from 10 to 1000 F $_\oplus$, in a timespan from 10 Myr up to 10 Gyr, considering a mass-loss efficiency of e = 0.1. Planets become bluer (less H/He mass in their atmospheres) and smaller as they lose their initial envelopes.

the planetary magnetic moment and the star-planet interactions, including the properties of the incident stellar winds (Kislyakova et al., 2014a; Vidotto and Bourrier, 2017; Khodachenko et al., 2015, 2017; Shaikhislamov et al., 2016).

Observation of the absorption produced by the planetary atmosphere in the wings of the Lyα line, however, offers a window to study the high velocity hydrogen atoms in the planetary atmosphere, as observed in the Neptune-like planet GJ436b (Ehrenreich et al. 2015; Kulow et al. 2014). The transmission spectrum in the Lyα line showed a clear absorption in the blue-wing of the spectrum, corresponding to velocities in the interval between −40 and −120 km/s. This asymmetric absorption in comparison to the more stable red-wing of the spectrum gives evidence of an extended, comet-like hydrogen cloud around the gaseous planet (see Fig. 2.27).

Hot hydrogen coronae could be formed around Earth-like exoplanets through the interaction between the stellar wind particles and the atmosphere of the planet. In charge-exchange reactions a fast proton of the stellar wind plasma interacts with an atmospheric neutral, the latter transferring an electron to the proton. An energetic

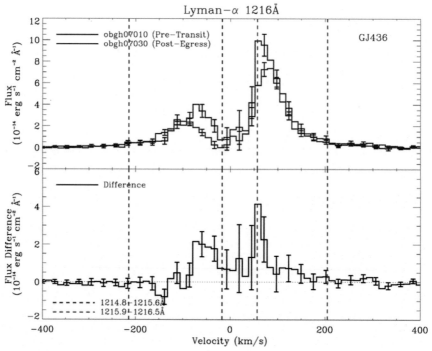

FIGURE 2.27

Evolution of the Ly-alpha emission line during the transit of GJ 436b, showing a clear asymmetric absorption from the planetary atmosphere. Top panel: Pre-ingress spectrum, in dark red, and post-egress spectrum, in dark blue; bottom panel: difference between these spectra, pre-ingress minus post-egress. The blue wing integrated flux region is between the blue dashed lines, and the red wing integrated region is between the red dashed lines.

From Kulow et al. (2014).

neutral atom (ENA) is then generated, having the same velocity of the initial stellar wind proton. As a consequence, hydrogen coronae of fast ENA may be present around rocky planets (Kislyakova et al., 2014b). The presence of such gaseous structure may contribute to the Lyα absorption in the wings of the line during the planetary transit.

4.2.1.2 Discrimination between biotic and abiotic routes for the detection of oxygen

As described in Section 2, the detection of O_2 in a planetary atmosphere does not lead unavoidably to the presence of live; abiotic chemical routes such as the photo-dissociation of CO_2 or H_2O may also result in a significant abundance of O_2. To discriminate between biotic and abiotic processes, e.g., to detect false positives, it is required to measure the absorption caused by ozone in the very sensitive Hartley band of ozone (Viallon et al., 2015), as well as the stellar UV radiation driving the photochemistry (Kasting, 1997; Leger et al., 1999).

False positives, may be caused by several reasons. Tian et al. (2014) and Harman et al. (2015) showed that planets orbiting M stars with high FUV to NUV ratio may increase their atmospheric O_2 and O_3 concentrations (see Fig. 2.28), leading to detectable amounts of these molecules, and consequently to false-positive detections. UV spectra of M dwarfs show strong emission in the FUV range, with net radiation comparable to or higher than the Sun, and mild NUV flux that could be three orders of magnitude lower than the Sun's (France et al., 2012a, 2013).

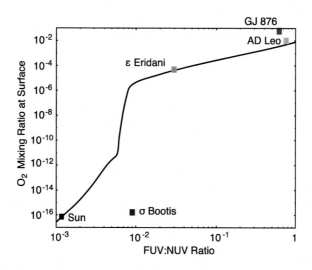

FIGURE 2.28

Dependence of O_2 mixing ratio (abundance of one component of a mixture relative to that of all other components) with the FUV/NUV ratio for the stars of the study of Harman et al. (2015).

O3 and O2 concentrations are also enhanced by stellar activity (Rugheimer et al., 2015) since the UV flux increases from the inactive to active stars and from the M9V to M0V stars (see Fig. 2.29). Loyd et al. (2016) predicted a similar trend, showing that the short-wavelength region of the optical band drives ozone photodissociation in low-NUV M-type stars, while for K stars, NUV radiation of the host star drives photolysis reactions of ozone.

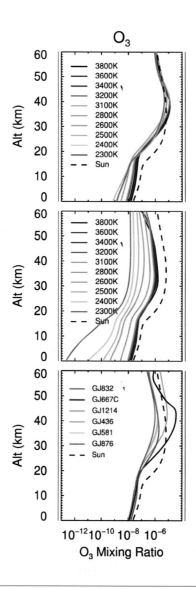

FIGURE 2.29

O_3 mixing ratio with altitude for an Earth-like planetary atmosphere, under the UV flux of M0 (T_{eff} = 3800K) to M9 (T_{eff} = 2300 K) stars. Upper panel: active stellar model. Middle panel: inactive stellar model. Bottom panel: MUSCLES program sample stars.

After Rugheimer et al. (2015).

Hot F-type stars also show high FUV fluxes. In a photochemistry model for the F2V star σ Boo including reactions with reduced radicals like NH_3, Domagal-Goldman et al. (2014) found that the abiotic column density of O_2 is a factor of 3000 below that of the present's day Earth, but the O_3 column density is only a factor of six smaller and could be detectable.

Besides the influence of the UV radiation of the parent star, the concentration of abiotic O_2 and O_3 is also strongly influenced by the presence of atmospheric reducing gases such as H_2 and CH_4 product of the geologic activity of the planet (Hu et al., 2012).

4.2.1.3 Stellar winds and radiation

Stellar winds drive nonthermal escape in planetary atmospheres. To properly evaluate the interaction of stellar wind plasma and upper atmospheric particles, an accurate characterization of these winds is required. UV observations provide the most successful method to measure stellar winds, through the observation of Lyα absorption in the stellar astrospheres (see Linsky contribution to this volume). Wood et al. (2005) reported the mass-loss rates obtained through this method for over a dozen of solar-like stars, showing a correlation with the stellar activity. Using the classical age-activity relation (Ayres, 1997), they found that mass-loss rate of solar-like stars could be inferred from the stellar age as:

$$\dot{M} \propto t^{-2.33 \pm 0.55}$$

However, this relation is not valid for the most active stars of the sample, including the young (500 Myr) solar-like star π Uma (Wood et al., 2014) and the 2 M stars of the sample, Proxima Cen, and EV Lac.

Winds of M dwarf stars are difficult to measure. Vidotto et al. (2011) derived the wind properties of V374Peg, a very active M star, showing that its wind produces a ram pressure (Pram = rho*v*v) five orders of magnitude higher than the Sun at 1 AU. The well-studied case of Proxima b showed enhanced levels of mass loss, showing a wind 40−80 times larger than seen at the present Earth (Ribas et al., 2016).

Photoionization by incident EUV and Lyα photons, along with collisions with stellar wind electrons and charge-exchange reactions driven by fast stellar wind protons, ionize the neutral particles present at the upper atmosphere of a planet. The magnetic field embedded in the stellar wind plasma may pick-up the resulting ions, consequently resulting in particle losses. These nonthermal losses may become important in weakly magnetized planetary environments, where the stellar wind penetrates deeper in the planetary atmosphere (Cohen et al., 2015), or by contrast, in highly irradiated atmospheres, expanded well-beyond the planet's magnetospheric protection. In the latter scenario, Kislyakova et al. (2013) studied the contribution of nonthermal processes to the total escape of H-rich, no-hydrostatic atmospheres in Earth-like and super-Earth-like planets under high levels of XUV radiation, concluding that nonthermal losses constitutes only a small fraction of the total planetary escape. This result is reproduced as well for the Super-Earths of the Kepler-11 planetary system (Kislyakova et al., 2014b), where nonthermal losses constitute a few percentages of the thermal losses predicted by Lammer et al. (2013), considering negligible magnetic moments, or for a nonmagnetized Venus-like planet in Cohen et al. (2015), showing that the nonthermal losses of the planet are insignificant during the planet's lifetime.

As described above, thermal losses of highly irradiated Earth-like planets will strip the H-rich protoatmospheres of these planets, resulting in oxygen- and nitrogen-rich atmospheres populated by heavier particles. These species are difficult to remove by thermal-escape mechanisms, while the presence of dense and fast stellar winds may contribute to their loss by nonthermal processes, see Fig. 2.30 (Canet and Gómez de Castro, 2021).

5. The chemical evolution and the missing metals problem

The strong interaction of UV radiation with matter makes of it an extremely useful, and in some very rarified environments the only, tool to study astrophysical processes. In the context of astrobiological research, UV observations enable the detection of the abundance of the elements needed to produce the biomolecules as well as to examine their distribution across space and along the history of the Universe. As illustrated in Fig. 2.31, the resonance transitions of the elements needed for life (H, C, O, P, S) are concentrated in the UV range, unless nitrogen.

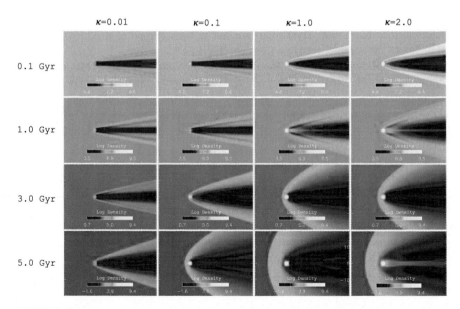

FIGURE 2.30

Evolution of the hydrogen exosphere of an Earth-like planet ($R = 1R_\oplus$) under different stellar wind conditions of stars from 0.1 to 5.0 Gyr, and different initial exospheric densities (from $\kappa = 0.01$ corresponding to a 1% of the Earth's exospheric density up to $\kappa = 2.0$, corresponding to a density twice of the Earth's).

After Canet and Gómez de Castro (2021).

All elements heavier than boron are produced within galaxies by stars and stellar byproducts such as supernovae explosions or massive post-AGB outflows. Hence, carbon (or silicon)-based life was not feasible in the very early Universe. There is not yet any precise determination of the time along the Universe history when live might have been initiated. Theoretical modeling and astronomical observations suggest that the abundance of important elements for biochemistry (C, N, O) was ~ 10 times smaller than today at $z \sim 4-5$ (see Vangioni et al., 2018 and references therein); at that time, the Universe was $\sim 1.1-1.6$ Gyr (Wright, 2006), about 10% of its current age: 13.721 Gyr. The low abundance of some key elements could have hampered significantly, the formation of Earth-like life at that time; in a sense, life is a result of the chemical evolution of the Universe. A simple example is posed by phosphor, the least abundant of the elements in Fig. 2.31. The abundance of phosphor in the solar system is 30.5 atoms per billion of hydrogen atoms, very small compared with that of O (6.76×10^5), N (1.18×10^5), C (3.71×10^5), and S (0.16×10^5), all given in the same units. Phosphor is believed to be mainly formed in massive stars in the pre-SN stage during the carbon and neon burning in a hydrostatic shell (Woosley and Weaver, 1995). A drop by one dex in the phosphor abundance could affect significantly to the formation of the phosphate groups that are part of adenosine triphosphate (ATP), RNA, DNA, and the cellular membranes (Gibard et al., 2018). The resonance transitions of P I and P II are in the 120–170 nm range

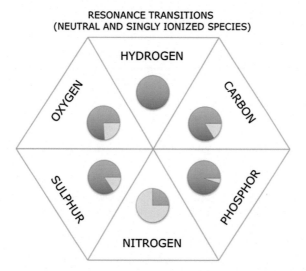

FIGURE 2.31

Fraction of resonance transitions of the basic bio-elements observed in the 90–320 nm spectral range; only the resonance transitions of neutral and singly ionized species are considered; the percent of transitions in the 90–320 nm is marked in blue; green otherwise.

and UV observations obtained with the Far Ultraviolet Spectroscopic Explorer (FUSE) and the Hubble Space Telescope have enabled the determination of phosphor abundance in the nearby ISM that it is found to be similar to the Solar one (Dufton et al., 1986, Lebouteiller and Ferlet, 2005). However, the abundance of highly ionized phosphor in the Cass A supernova remnant is about 100 times higher than in the ISM (Koo et al., 2013). Hence mixing processes between afresh, enriched material and the underlying ISM must play an important role in the average chemical abundance in any area in space.

In general, there is a missing metals problem at galactic scale. Only $\sim 20\%$ of all metals produced in galaxies are retained within. The metal mass densities inferred from damped Lyα systems at $z \in [0.5, 5]$ are a factor of 10 lower than expected from the star formation history of the universe (Prochaska et al., 2003). According to calculations, roughly $80\%-85\%$ of the metals (by mass) are produced in core-collapse supernovae; the rest is produced in Type Ia supernovae explosions and by post-AGB stars. However, the abundances predicted by the theory exceed by a factor of ~ 4 the total inventory of metals, including metals trapped in stars and those released to the ISM either in the form of gas or dust (Peeples et al., 2014).

Indeed, metals are transported away from the originating galaxy into the intergalactic medium by supernova-driven winds over periods of several Gyr. There is also a significant loss by the effect of radiation pressure on dust grains (Shustov and Vibe, 1995; Sharma et al., 2011). This impoverishment takes place during the whole life of galaxies and results in the formation of extended galactic halos with strong radial metallicity gradients (Crain et al., 2013). Critical tests to these models are run at UV wavelengths for $z < 1.5$ galaxies by measuring the physical extent of metals around galaxies, and their relative mixtures in the phases of the gas. Galactic metallicity can also be reduced by the contribution from fresh, low metallicity intergalactic material accreting onto the galaxies (Martin et al., 2019). Thus, to calibrate the influence of galactic-scale processes in the global chemical enrichment, it is required to address the interaction of galaxies with the intergalactic medium and hence, galaxy formation and evolution.

At small scale, within any given galaxy, there may be metal enriched niches meeting the conditions for biomolecules to be formed. These niches are to be located preferentially, in the centers of galaxies like ours, at the locations where fresh galactic and intergalactic gas is preferentially dragged into by the gravitational field driving at efficient star formation and fast chemical evolution (see Fig. 2.32).

6. The UV path to the detection of amino acids in space

Investigations concerning generation of life at a cosmic level have been severely hampered by difficulties with the collection of experimental data and remote detection of even simple compounds like amino acids. To date, only glycine has been conclusively detected using *in situ* measurements carried out by the Rosetta probe (Altwegg et al., 2016). However, remote detection could be possible if the chirality imbalance observed in many biological molecules on Earth were widespread in the Universe. According to laboratory experiments, in space ice α-alanine (hereafter,

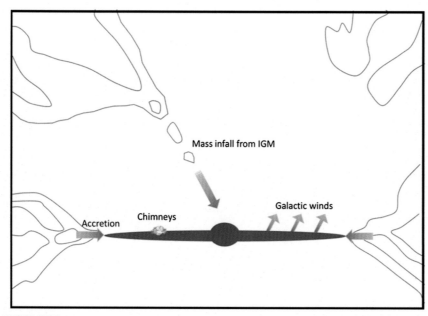

FIGURE 2.32

Cartoon illustrating the main processes involved in the chemical evolution of galaxies.

alanine) is the second-most abundant amino acid after glycine, with an abundance 40% that of glycine (Muñoz-Caro et al., 2002). Unlike glycine, alanine is an optically active molecule. Left and right isomers of alanine are produced naturally and any imbalance in isomer distribution will produce a clear signature in the polarization of the radiation.

The structural asymmetry of alanine results in different interactions with left-handed (L) and right-handed (R) circularly polarized radiation (CPR). The most prominent difference can be observed at 180 nm and is usually ascribed to the $\pi_0 \rightarrow \pi^*$ transitions of the COO- group (Kaneko et al., 2009). This optical activity is manifested through two related phenomena: optical rotatory dispersion (ORD) and circular dichroism (CD) that produce specific features in the polarization of the UV spectrum especially in space ices (comets, interstellar dust, planetary caps …). The UV spectrum of comets has a prominent S I feature at 180 nm that might made feasible the detection of the alanine imprint in the polarization spectrum (see Fig. 2.33).

Radiation from comets is known to be polarized both linearly and circularly (Hines and Levasseur-Regourd, 2016; (Rosenbush et al. (2007)). Circular polarization of ∼1% has been detected at optical wavelengths that could be produced by several mechanisms, including multiple scattering in asymmetric particle distributions and scattering by intrinsically asymmetric particles that may (Nagdimunov et al., 2013) or may not (Guirado et al., 2007) be organic. However, the CD and ORD signature at UV wavelengths is unique to amino acids. Unfortunately, UV spectropolarimetric observations of comets have not been carried out yet.

The expected CD and ORD signal from comets at UV wavelengths has been evaluated recently (Gómez de Castro and de Isidro-Gómez, 2021). Estimates have

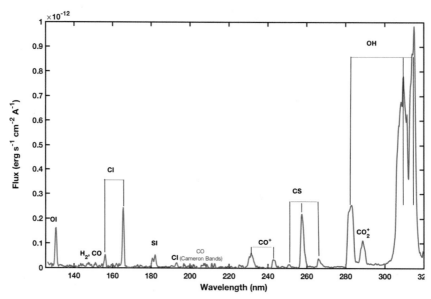

FIGURE 2.33

UV spectrum of comet Bradley as observed by the International Ultraviolet Explorer (IUE) in 1980 (Weaver et al., 1981). The main spectral features are marked, including the location of the Cameron CO bands that are not observed in IUE data. The OH feature is saturated in the image. The continuum produced by comet-induced solar radiation scattering is very weak. The comet was at heliocentric distance 0.71 AU and 0.61 AU from the IUE (Feldman et al., 1980).

been made using the alanine to glycine scaling derived from laboratory experiments (Muñoz-Caro et al., 2002), $n_{ala}/n_{gly} \sim 0.4$, and the measurement of the glycine abundance in the tail of the comet 67P/Churyumov-Gerasimenko by the ROSINA mass spectrometer in the Rosetta probe (Altwegg et al., 2016). ROSINA detected an abundance of glycine of 170 ppb in the gas phase, released from sublimation of the icy mantles of dust grains in the coma (Hadraoui et al., 2019).

Alanine's ORD introduces a wavelength-dependent additional rotation to the linearly polarized radiation from comets. The linear polarization at optical wavelengths may be as high as 25% (especially at large phase angles) and decreases toward the UV (Hines and Levasseur-Regourd, 2016); an extrapolation from optical data (Kiselev et al., 2008) results in expected values of about 7% at 180 nm for reasonable phase angles of $40°-60°$. The added rotation angle by alanine's optical activity, α_{ORD}, is,

$$\alpha_{ORD}(arcmin) \cong 2.5280 \left(\frac{Z}{2 \times 10^6 cm}\right) \left(\frac{n}{8.4. \times 10^6 cm^{-3}}\right) \left(\frac{I^l(\lambda)}{200}\right) ee$$

where Z and n are the thickness and density of the glycine layer determined for comet 67P/Churyumov-Gerasimenko from ROSINA measurements (Hadraoui et al., 2019). The observed signal depends on the enantiomeric excess (*ee*), e.g., the excess density of the levo isomer (L-alanine) with respect to the dextro isomer

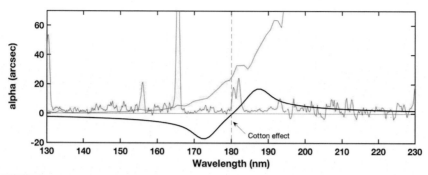

FIGURE 2.34

α_{obs} for $ee = 0.1$(black). Comet Bradley (blue) and the solar spectrum (red) are shown, in arbitrary units, for reference.

(r-alanine). Enantiomeric excesses in meteorites are in the range 1%−10% (Pizzarello and Croning, 2000). α_{obs} is displayed in Fig. 2.34 for an $ee = 0.1$ and compared with the spectrum of comet Bradley and the solar spectrum.

Calculations based on a simple instrumental configuration, show that measuring the ORD is achievable for $ee = 1$ even for small, 50 cm primary, observatories and the low density of 67P/Churyumov-Gerasimenko coma. However, reaching more realistic $ee = 0.1$ is only achievable for medium size (3−4 m primary) space telescopes and for the brightest comets (Halley, Levy …). Detecting alanine in samples with $ee \leq 0.01$ is not feasible by remote observation however, small spectropolarimeters in exploration missions (such as Rosetta) are an efficient mean to remote sense the presence of alanine in comets and planetary ices (i.e., Europa) without requiring the procurement of any sample and the associated contamination issues.

Remote sensing of alanine in the Solar System will not only enable studying the distribution and abundance of amino acids, but also investigating the source of the enantiomeric imbalance of fundamental molecules for life (amino acids, sugars) on Earth. The majority of amino acids existing on Earth have a pre-defined chirality whose origin is poorly understood (Giri et al., 2013). Measuring whether L-alanine in comets is the most abundant isomer, as it is on Earth, is a challenging, but feasible task at UV wavelengths.

7. Summary and conclusions

This review provides a necessarily short compilation of the most important processes and methods concerning UV astronomy and the investigation of the origin of life. As we show, UV astronomy provides some unique tools to address this research. Also, UV radiation from astronomical sources drives important processes and needs to be known for the robust appraisal of relevant phenomena observed at other wavelengths (see Table 2.3 for a summary). It is this dual character what makes of UV observatories instrumental for the astronomical investigation of the origin of life.

Table 2.3 Summary.

Relevant phenomena	Main tracers (X-ray: XR; ultraviolet: UV; visible: V; infrared: IR; radio wavelengths: RW)
Chemical evolution (z < 1.8)	
Star formation efficiency and initial mass function (also in extended disks and dwarf galaxies)	Parent molecular clouds and embedded populations (IR, RW) PDRs (IR, RWs) H II regions (UV, V) Massive stars (UV) High sensitivity to star formation and extended disks (UV)
Galactic winds and chimneys Accretion of extragalactic material	Chemical composition of gas at temperatures above 1 MK (XR) Chemical composition and dynamics of gas at temperatures between 1,000 to 1 MK (UV); best sensitivity Chemical composition and dynamics of gas at temperatures from $\sim 10^3$ to 10^4 K (V).
Intergalactic medium/ Cosmic web	Chemical enrichment in the voids of galaxies (XR) Chemical enrichment in filaments and galactic halos (UV)
Post-AGB stars and SN shells	Winds and outflows (UV, V, IR) Shells (XR, UV)
Formation and evolution of planetary systems	
Accretion physics	Disk heating by the accretion flow (IR, RW) Accretion shocks (UV, XR) Dynamical processes in accretion shocks (UV, V) Inflated magnetospheres (UV, V)
Jets and outflows	Embedded jets and molecular outflows (IR, RW) HH objects and low excitation jets (V, UV) Base of winds (disk-winds) (UV, V, IR)
Proto planetary disk evolution	Young disks and chemical evolution (IR, RW) Dust distribution (IR, RW) Inner (<1 AU) disk (UV) Photoevaporative flows (UV, V, IR) Debris disks (UV, V, IR)
Magnetic activity	Stellar photosphere (V, IR) Chromosphere (high sensitivity UV but also V (Ca II, Hα)) Transition region (UV) Corona (UV, XR) Flares (XR, UV, V)

Table 2.3 Summary.—*cont'd*

Relevant phenomena	Main tracers (X-ray: XR; ultraviolet: UV; visible: V; infrared: IR; radio wavelengths: RW)
Exoplanets: exoplanetary systems	
Planetary atmospheres and winds	Low atmosphere, clouds, and weather patterns (V, IR) Ozone, bio-tracer, and false positive detection (UV) High atmosphere and planetary winds (UV, V)
Stellar winds and radiation	Photoionizing radiation field (UV) Mass loss rates (UV) Stellar activity (V, UV, XR)
Detection of amino acids and proteins	
Detection of chiral amino acids	ORD and CD signature of abundant amino acids (UV)
Nucleobases absorbance	UV

There are several UV observatories currently under development for the investigation of these processes and ambitious projects to develop the large facilities that will enable the observation of the atmospheres of Earth-like planets further than the Solar System to seek for evidence of life. A summary on the current status of these missions and projects is provided in the last chapter of the volume.

Acknowledgments

The authors want to thank Noah Brosch and Boris Shustov for the revision of the text. The work has been partially funded by the Ministry of Science and Innovation of Spain through grant: ESP2017-87813-R.

References

Abdu, Y.A., Hawthorne, F.C., Varela, M.E., 2018. Astrophys. J. 856, L9.

Agúndez, M., Cernicharo, J., Quintana-Lacaci, G., et al., 2017. A&A 601, A4.

Alexander, R., Pascucci, I., Andrews, S., et al., 2014. In: Beuther, H., Klessen, R.S., Dullemond, C.P., Henning, T. (Eds.), Protostars and Planets VI. University of Arizona Press, Tucson, p. 475, 914 pp.

ALMA Partnership, 2015. Astrophys. J. Lett. 808, L3.

Altwegg, K., Balsiger, H., Bar-Nun, A., et al., 2016. Sci. Adv. 2, e1600285.

Anderson, K.R., Adams, F.C., Calvet, N., 2013. Astrophys. J. 774, 49.

Anglada-Escudé, G., Amado, P.J., Barnes, J., et al., 2016. Nature 536, 437.

Ardila, D.R., Herczeg, G.J., Gregory, S.G., et al., 2013. ApSS 207, 1.

Armitage, P.J., 2007. Astrophys. J. 665, 1381.

Armitage, P.J., Livio, M., Lubow, S.H., et al., 2002. MNRAS 334, 248.

Arney, G., Domagal-Goldman, S.D., Meadows, V.S., et al., 2016. Astrobiology 16.

Arney, G., Domagal-Goldman, S.D., Meadows, V.S., 2018. Astrobiology 18.

Ayres, T.R., Fleming, T.A., Simon, T., et al., 1995. Astrophys. J. Supp. 96, 223.

Ayres, T.R., 1997. JGR 102, 1641.

Bai, X.-N., 2017. Astrophys. J. 845, 75.

Balbus, S.A., Hawley, J.F., 1991. Astrophys. J. 376, 214.

Balbus, S.A., Papaloizou, J.C.B., 1999. Astrophys. J. 521, 650.

Baliukin, I.I., Bertaux, J.-L., Quémerais, E., et al., 2019. JCR Space Phys. 124, 861.

Banzatti, A., Pinilla, P., Ricci, L., et al., 2015. Astrophys. J. 815, L15.

Bergin, E., Calvet, N., D'Alessio, P., et al., 2003. Astrophys. J. 591, L159.

Bergin, E.A., Cleeves, L.I., Gorti, U., et al., 2013. Nature 493, 644.

Bernstein, M.P., Sandford, S.A., Allamandola, L.J., et al., 1995. Astrophys. J. 454, 327.

Bernstein, M.P., Dworkin, J.P., Sandford, S.A., et al., 2002. Nature 416, 401−403.

Beth, A., Garnier, P., Toublanc, D., et al., 2016. Icarus 266, 410.

Bethell, T.J., Bergin, E.A., 2011. Astrophys. J. 739, 78.

Bishop, J., Chamberlain, J.W., 1989. Icarus 81, 145.

Bolmont, E., Selsis, F., Owen, J.E., et al., 2017. MNRAS 464, 3728.

Bourrier, V., Ehrenreich, D., King, G., et al., 2017a. A&A 597, A26.

Bourrier, V., Ehrenreich, D., Allart, R., et al., 2017b. A&A 602, A106.

Bourrier, V., Lecavelier des Etangs, A., Ehrenreich, D., et al., 2018. A&A 620, A147.

Brasseur, G., Solomon, S., 1996. Aeronomy of the middle atmosphere. In: Chemistry and Physics of the Stratosphere and Mesosphere, Springer ed., vol. 5. Atmospheric Science Library, ISBN 978-90-277-2344-4.

Brown, L.R., Troutman, M.R., Gibb, E.L., 2013. Astrophys. J. 770, L14.

Caballero, J., 2007. A&A 466, 917.

Camprubí, E., de Leeuw, J.W., House, C.H., et al., 2019. SS Rev. 215, 56.

Canet, A., Gómez de Castro, A.I., 2021. MNRAS. submitted.

Carr, J.S., Najita, J.R., 2011. Astrophys. J. 733, 102.

Carruthers, G.R., Page, T., 1972. Science 177, 788.

Carruthers, G.R., 1973. Appl. Optic. 12, 2501.

Catling, D.C., Zahnle, K.J., McKay, C.P., 2001. Science 293, 839.

Catling, D.C., Krissansen-Totton, J., Kiang, N.Y., et al., 2018. Astrobiology 18, 709.

Chamberlain, J.W., 1963. Planet. Space Sci. 11, 901.

Chamberlain, J.W., 1979. Icarus 39, 286.

Chapman, S., 1930. R. Meteorol. Soc. Memoir. 3, 103.

Charbonneau, D., Berta, Z.K., Irwin, J., et al., 2009. Nature 462, 891.

Chen, Y.-J., Nuevo, M., Yih, T.-S., et al., 2008. MNRAS 384, 605.

Christensen, P.R., Pearl, J.C., 1997. JGR 102, 10875.

Claire, M.W., Catling, D.C., Zahnle, K.J., 2006. Geobiology 4, 239.

Clarke, C.J., Gendrin, A., Sotomayor, M., 2001. MNRAS 328, 485.

Cleaves, H.J., Chalmers, J.H., Lazcano, A., et al., 2008. Orig. Life Evol. Biosph. 38, 105.

Cockell, C.S., Bush, T., Bryce, C., et al., 2016. Astrobiology 16, 89.

Cockell, C.S., 2000. Orig. Life Evol. Biosph. 30, 467.

Coffey, D., Dougados, C., Cabrit, S., et al., 2015. Astrophys. J. 804, 2.

Cohen, O., Ma, Y., Drake, J.J., et al., 2015. Astrophys. J. 806, 41.

Collins, J.A., Shull, J., Giroux, M.L., 2009. Astrophys. J. 705, 962.

Crain, R.A., McCarthy, I.G., Schaye, J., et al., 2013. MNRAS 432, 3005.

de Marcellus, P., Bertrand, M., Nuevo, M., et al., 2011. Astrobiology 11, 847.

Des Marais, D.J., Harwit, M.O., Jucks, K.W., et al., 2002. Astrobiology 2, 153.

Dessler, A.E., Weinstock, E.M., Hintsa, E.J., et al., 1994. GRL 21, 2563.

Dipierro, G., Price, D., Laibe, G., et al., 2015. MNRAS 453, 73.

Domagal-Goldman, S.D., Kasting, J.F., Johnston, D.T., 2008. Earth Planet Sci. Lett. 269, 29.

Domagal-Goldman, S.D., Segura, A., Claire, M.W., et al., 2014. Astrophys. J. 792, 90.

Donati, J.-F., Landstreet, J.D., 2009. ARAA 47, 333.

Dos Santos, L.A., Bourrier, V., Ehrenreich, D., et al., 2019. A&A 622, A46.

Draine, B.T., 2016. Astrophys. J. 831, 109.

Dressing, C.D., Charbonneau, D., 2015. Astrophys. J. 807, 45.

Dufton, P.L., Keenan, F.P., Hibbert, A., 1986. A&A 164, 179.

Ehling-Shultz, M., Bilger, W., Scherer, S., 1997. J. Bacteriol. 179, 1940.

Ehrenreich, D., Lecavelier Des Etangs, A., Hébrard, G., et al., 2008. A&A 483, 933.

Ehrenreich, D., Bourrier, V., Bonfils, X., et al., 2012. A&A 547, A18.

Ehrenreich, D., Bourrier, V., Wheatley, P.J., et al., 2015. Nature 522, 459.

Ercolano, B., Clarke, C.J., Drake, J.J., 2009. Astrophys. J. 699, 1639.

Erkaev, N.V., Lammer, H., Odert, P., et al., 2013. Astrobiology 13, 1011.

Erkaev, N.V., Lammer, H., Odert, P., et al., 2016. MNRAS 460, 1300.

Feldman, P.D., Weaver, H.A., Festou, M., et al., 1980. Nature 286, 132.

Feng, Y.K., Robinson, T.D., Fortney, J.J., et al., 2018. Astrophys. J. 155, 200.

Fogel, J.K.J., Bethell, T.J., Bergin, E.A., et al., 2011. Astrophys. J. 726, 29.

Font, A.S., McCarthy, I.G., Johnstone, D., et al., 2004. Astrophys. J. 607, 890.

Fossati, L., Haswell, C.A., Froning, C.S., et al., 2010. Astrophys. J. 714, L222.

France, K., Linsky, J.L., Brown, A., et al., 2010. Astrophys. J. 715, 596.

France, K., Schindhelm, E., Burgh, E.B., et al., 2011. Astrophys. J. 734, 31.

France, K., Linsky, J.L., Tian, F., et al., 2012a. Astrophys. J. 750, L32.

France, K., Schindhelm, E., Herczeg, G.J., et al., 2012b. Astrophys. J. 756, 171.

France, K., Froning, C.S., Linsky, J.L., et al., 2013. Astrophys. J. 763, 149.

Frondel, C., Marvin, U.B., 1967. Nature 217, 587.

Fukuzawa, K., Osamura, Y., Schaefer, H.F., 1998. Astrophys. J. 505, 278.

Fuselier, S.A., Burch, J.L., Lewis, W.S., et al., 2000. SSRev 91, 51.

Gammie, C.F., 2001. Astrophys. J. 553, 174.

Garcia-Pichel, F.A., 1994. Limnol. Oceanogr. 39, 1704.

Gerakines, P.A., Schutte, W.A., Ehrenfreund, P., 1996. A&A 312, 289.

Gerakines, P.A., Moore, M.H., Hudson, R.L., 2001. JGR 106, 33381.

Gerakines, P.A., Moore, M.H., Hudson, R.L., 2004. Icarus 170, 202.

Gibard, C., Bhowmik, S., Karki, E., et al., 2018. Nat. Chem. 10, 212.

Gillon, M., Triaud, A.H.M.J., Demory, B.-O., et al., 2017. Nature 542, 456.

Giri, C., Goesmann, F., Meinert, C., et al., 2013. Top Curr. Chem. 333, 41.

Goldblatt, C., Watson, A.J., Lento, T.M., 2009. Bioastronomy 2007: Molecules, p. 420.

Gómez de Castro, A.I., Lamzin, S.A., 1999. MNRAS 304, L41.

Gómez de Castro, A.I., Robles, A., 1999. INES Access Guide No. 1: Herbig-Haro Objects. ESA Publications.

Gómez de Castro, A.I., Verdugo, E., 2001. Astrophys. J. 548, 976.

Gómez de Castro, A.I., 2002. MNRAS 332, 409.

Gómez de Castro, A.I., Verdugo, E., 2003. Astrophys. J. 597, 443.

Gómez de Castro, A.I., Ferro-Fontán, C., 2005. MNRAS 362, 569.

Gómez de Castro, A.I., von Rekowski, B., 2011. MNRAS 411, 849.

Gómez de Castro, A.I., Marcos-Arenal, P., 2012. Astrophys. J. 749, 190.

Gómez de Castro, A.I., 2013a. In: Oswalt, T.D., Barstow, M.A. (Eds.), Planets, Stars and Stellar Systems, vol. 4. Springer Science+Business Media Dordrecht, ISBN 978-94-007-5614-4, p. 279.

Gómez de Castro, A.I., 2013b. Astrophys. J. 775, 131.

Gómez de Castro, A.I., Ustamujic, S., 2015. Accretion and Outflow Workshop. Noordwijk. https://www.cosmos.esa.int/web/accretion-outflow-workshop/program.

Gómez de Castro, A.I., Beitia-Antero, L., Ustamujic, S., 2018. ExpA 45, 147.

Gómez de Castro, A.I., Vallejo, J.C., Canet, A., et al., 2020. Astrophys. J. 904, 120.

Gómez de Castro, A.I., Rheinstadter, M., Clancy, P., et al., 2021. Sci. Rep. 11, ArtNo. 2492

Gómez de Castro, A.I., de Isidro-Gómez, A.I., 2021. Astrobiology, 21, Issue 7

Goodson, A.P., Winglee, R.M., Böhm, K.-H., 1997. Astrophys. J. 489, 199.

Gorti, U., Dullemond, C.P., Hollenbach, D., 2009. Astrophys. J. 705, 1237.

Gorti, U., Hollenbach, D., 2009. Astrophys. J. 690, 1539.

Gorti, U., Hollenbach, D., Dullemond, C.P., 2015. Astrophys. J. 804, 29.

Greaves, J.S., Scaife, A.M.M., Frayer, D.T., et al., 2018. Nat. Astron 2, 662.

Grenfell, J.L., 2017. Phys. Rep. 713, 1.

Grim, R.J.A., Greenberg, J.M., 1987. Astrophys. J. 321, L91.

Guirado, D., Hovenier, J.W., Moreno, F., 2007. JQS&RT 106, 63.

Gullbring, E., Calvet, N., Muzerolle, J., et al., 2000. Astrophys. J. 544, 927.

Hadden, S., Lithwick, Y., 2014. Astrophys. J. 787, 80.

Hadraoui, K., Cottin, H., Ivanovski, S.L., et al., 2019. A&A 630, A32.

Hanneman, R.E., Strong, H.M., Bundy, F.P., 1967. Science 155, 995.

Hanslmeier, A., Kempe, S., Seckbach, J., 2012. In: Life on Earth and Other Planetary Bodies: Cellular Origin, Life in Extreme Habitats and Astrobiology, vol. 24. Springer Science + Business Media, Dordrecht, ISBN 978-94-007-4965-8.

Haqq-Misra, J.D., Domagal-Goldman, S.D., Kasting, P.J., et al., 2008. Astrobiology 8, 1127.

Harman, C.E., Schwieterman, E.W., Schottelkotte, J.C., et al., 2015. Astrophys. J. 812, 137.

Hartmann, L., Herczeg, G., Calvet, N., 2016. ARA&A 54, 135.

Hasegawa, T.I., Herbst, E., Leung, C.M., 1992. ApJSS 82, 167.

Hasegawa, T.I., Herbst, E., 1993. Mon. Notices Royal Astron. Soc. 263, 589.

Haswell, C.A., Fossati, L., Ayres, T., et al., 2012. Astrophys. J. 760, 79.

Hendrix, A.R., Retherford, K.D., Gladstone, G.R., et al., 2012. JGR (Planets) 117, E12001.

Herbst, E., 1988. In: Millar, T.J., Williams, D.A. (Eds.), Rate Coefficients in Astrochemistry. Kluwer Academic Publishers, Dordrecht, pp. 239−262.

Hernández, J., Calvet, N., Briceño, C., et al., 2007. Astrophys. J. 671, 1784.

Hijnen, W.A.M., Beerendok, E.F., Medema, G.J., 2006. Water Res. 40, 3.

Hines, D.C., Levasseur-Regourd, A.-C., 2016. P&SS 123, 41.

Holland, H.D., 2002. Geochem. Cosmochim. Acta 66, 3811−3826.

Hollenbach, D.J., Yorke, H.W., Johnstone, D., 2000. In: Mannings, V., Boss, A.P., Russell, S.S. (Eds.), Protostars and Planets IV. University of Arizona Press, Tucson, pp. 401−428.

Hollenbach, D., Johnstone, D., Lizano, S., et al., 1994. Astrophys. J. 428, 654.

Hörneck, G., Rettberg, P., Reitz, G., et al., 2001a. Orig. Life Evol. Biosph. 31, 527.

Hörneck, G., Stöffler, D., Eschweiler, U., et al., 2001b. Icarus 149, 285.

Howard, A.W., Marcy, G.W., Bryson, S.T., et al., 2012. ApJSS 201, 15.

Hu, R., Seager, S., Bains, W., 2012. Astrophys. J. 761, 166.

Hu, R., Seager, S., Bains, W., 2013. Astrophys. J. 769, 6.

Hurley, J., Irwin, P.G.J., Adriani, A., et al., 2014. P&SS 90, 37.

Huss, G.R., Lewis, R.S., 1994. Meteoritics 29, 791.

Huss, G.R., Meshik, A.P., Smith, J.B., et al., 2003. Geochem. Cosmochim. Acta 67, 4823.

Iglesias-Groth, S., 2004. Astrophys. J. 608, L37.

Jenkins, E.B., Jura, M., Loewenstein, M., 1983. Astrophys. J. 270, 88.

Jiménez-Serra, I., Martín-Pintado, J., Rivilla, V.M., et al., 2020. Astrobiology 20, 1048.

Jin, S., Mordasini, C., Parmentier, V., et al., 2014. Astrophys. J. 795, 65.

Joblin, C., Léger, A., Martin, P., 1992. Astrophys. J. 393, L79.

Johns-Krull, C.M., Valenti, J.A., Linsky, J.L., 2000. Astrophys. J. 539, 815.

Johnstone, C.P., Güdel, M., Stökl, A., et al., 2015. Astrophys. J. 815, L12.

Jönsson, K.I., Rabbow, E., Schill, R.O., et al., 2008. Curr. Biol. 18, R729.

Kaltenegger, L., Traub, W.A., Jucks, K.W., 2007. Astrophys. J. 658, 598.

Kaltenegger, L., Sasselov, D., 2010. Astrophys. J. 708, 1162.

Kaltenegger, L., 2017. ARA&A 55, 433–485.

Kama, M., Bruderer, S., Carney, M., et al., 2016. A&A 588, A108.

Kameda, S., Ikezawa, S., Sato, M., et al., 2017. Geophys. Res. Lett. 44 (11), 706.

Kaneko, F., Yagi-Watanabe, K., Tanaka, M., et al., 2009. JPSJ 78, 013001.

Kasting, J.F., 1997. Orig. Life Evol. Biosph. 27, 291.

Kasting, J.F., Donahue, T.M., 1980. JGR 85, 3255.

Kasting, J.F., Whitmire, D.P., Reynolds, R.T., 1993. Icarus 101, 108.

Khodachenko, M.L., Shaikhislamov, I.F., Lammer, H., et al., 2015. Astrophys. J. 813, 50.

Khodachenko, M.L., Shaikhislamov, I.F., Lammer, H., et al., 2017. Astrophys. J. 847, 126.

Kiang, N.Y., Domagal-Goldman, S., Parenteau, M.N., et al., 2018. Astrobiology 18, 619.

Kim, S.J., A'Hearn, M.F., 1991. Icarus 90, 79.

Kiselev, N., Rosenbush, V., Kolokolova, L., et al., 2008. JQS&RT 109, 1384.

Kislyakova, K.G., Lammer, H., Holmström, M., et al., 2013. Astrobiology 13, 1030.

Kislyakova, K.G., Holmström, M., Lammer, H., et al., 2014a. Science 346, 981.

Kislyakova, K.G., Johnstone, C.P., Odert, P., et al., 2014b. A&A 562, A116.

Kislyakova, K.G., Holmström, M., Odert, P., et al., 2019. A&A 623, A131.

Kitadai, N., Maruyama, S., 2018. Geosci. Front. 9, 1117.

Koenigl, A., 1991. Astrophys. J. 370, L39.

Koo, B.-C., Lee, Y.-H., Moon, D.-S., et al., 2013. Science 342 (6164), 1346.

Kopparapu, R.K., Ramirez, R., Kasting, J.F., et al., 2013. Astrophys. J. 765, 131.

Kubyshkina, D., Lendl, M., Fossati, L., et al., 2018. A&A 612, A25.

Kulow, J.R., France, K., Linsky, J., et al., 2014. Astrophys. J. 786, 132.

Kump, L.R., 2008. Nature 451, 277.

Lammer, H., Kasting, J.F., Chassefière, E., et al., 2008. Space Sci. Rev. 139, 399.

Lammer, H., Bredehöft, J.H., Coustenis, A., et al., 2009. A&AR 17, 181.

Lammer, H., Erkaev, N.V., Odert, P., et al., 2013. MNRAS 430, 1247.

Lammer, H., 2013. Origin and Evolution of Planetary Atmospheres: Implications for Habitability. SpringerBriefs in Astronomy, ISBN 978-3-642-32086-6.

Lammer, H., Stökl, A., Erkaev, N.V., et al., 2014. MNRAS 439, 3225.

Lamzin, S.A., 1998. ARep 42, 322.

Lebouteiller, V.K., Ferlet, R., 2005. A&A 443, 509.

Lecavelier des Etangs, A., Vidal-Madjar, A., Roberge, A., et al., 2001. Nature 412, 706.

Lecavelier des Etangs, A., Deleuil, M., Vidal-Madjar, A., et al., 2003. A&A 407, 935.

Lecavelier Des Etangs, A., Ehrenreich, D., Vidal-Madjar, A., et al., 2010. A&A 514, A72.
Lecavelier des Etangs, A., Bourrier, V., Wheatley, P.J., et al., 2012. A&A 543, L4.
Léger, A., Ollivier, M., Altwegg, K., et al., 1999. A&A 341, 304.
Lewis, R.S., Ming, T., Wacker, J.F., et al., 1987. Nature 326, 160.
Lin, D.N.C., Hayashi, M., Bell, K.R., et al., 1994. Astrophys. J. 435, 821.
Lin, D.N.C., Papaloizou, J.C.B., 1996. ARA&A 34, 703.
Linsky, J., 2019. Lecture Notes in Physics. Berlin Springer-Verlag, p. 955.
Linsky, J.L., Yang, H., France, K., et al., 2010. Astrophys. J. 717, 1291.
Liseau, R., Huldtgren, M., Fridlund, C.V.M., et al., 1996. A&A 306, 255.
Lisman, D., Schwieterman, E., Reinhard, C., et al., 2019. Bul. AAS 51, 225.
Lissauer, J.J., Fabrycky, D.C., Ford, E.B., et al., 2011. Nature 470, 53.
Livengood, T.A., Deming, L.D., A'Hearn, M.F., et al., 2011. Astrobiology 11, 907.
Lopez, E.D., Fortney, J.J., Miller, N., 2012. Astrophys. J. 761, 59.
López-Martínez, F., Gómez de Castro, A.I., 2014. MNRAS 442, 2951.
López-Martínez, F., Gómez de Castro, A.I., 2015. MNRAS 448, 484.
Lopez, E.D., 2017. MNRAS 472, 245.
Loyd, R.O.P., France, K., Youngblood, A., et al., 2016. Astrophys. J. 824, 102.
Luger, R., Barnes, R., 2015. Astrobiology 15, 119.
Lundkvist, M.S., Kjeldsen, H., Albrecht, S., et al., 2016. Nat. Commun. 7, 11201.
Luo, G., Ono, S., Beukes, N.J., et al., 2016. Sci. Adv. 2, e1600134.
Lyons, T., Reinhard, C., Planavsky, N., 2014. Nature 506, 307.
Malloci, G., Mulas, G., Cecchi-Pestellini, C., et al., 2008. A&A 489, 1183.
Martin, C.L., Ho, S.H., Kacprzak, G.G., et al., 2019. Astrophys. J. 878, 84.
Martins, F., Mahy, L., Hillier, D.J., et al., 2012. A&A 538, A39.
Materese, C.K., Nuevo, M., Sandford, S.A., 2017. Astrobiology 17, 761−770.
McClure, M.K., Bergin, E.A., Cleeves, L.I., et al., 2016. Astrophys. J. 831, 167.
McComas, D.J., Allegrini, F., Baldonado, J., et al., 2009. SSRev 142, 157.
McElroy, D., Walsh, C., Markwick, A.J., et al., 2013. A&A 550, A36.
Meadows, V.S., 2017. Astrobiology 17, 1022.
Meftah, M., Damé, L., Bolsée, D., et al., 2020. Sol. Phys. 295, 14.
Meinert, C., Hoffmann, S.V., Cassam-Chenai, P., et al., 2014. Angew. Chem. Int. Ed. 53, 210.
Mende, S.B., Heetderks, H., Frey, H.U., et al., 2000. SSRev 91, 287.
Mihalas, D., 1978. Stellar Atmospheres. W.H. Freeman, San Francisco.
Mileikowsky, C., Cucinotta, F.A., Wilson, J.W., et al., 2000. Icarus 145, 391.
Miles, B.E., Roberge, A., Welsh, B., 2016. Astrophys. J. 824, 126.
Miller, S.L., 1953. Science 117 (3046), 528−529.
Miller, S.L., Urey, H.C., 1959. Science 130 (3370), 245−251.
Misra, A., Krissansen-Totton, J., Koehler, M.C., et al., 2015. Astrobiology 15, 462.
Mittler, R., Tel-Or, E., 1991. Free Radic. Res. Commun. 12, 845.
Mlozewska, A.M., Cole, D.V., Planavsky, M.J., et al., 2018. Nat. Commun. 9, 3088.
Muñoz Caro, G.M., Meierhenrich, U.J., Schutte, W.A., et al., 2002. Nature 416, 403.
Nagdimunov, L., Kolokolova, L., Mackowski, D., 2013. JQS&RT 131, 59.
Németh, P., Garvie, L.A.J., Aoki, T., et al., 2014. Nat. Commun. 5, 5447.
Nicholson, W.L., Munakata, N., Horneck, G., et al., 2000. Microbiol. Mol. Biol. Rev. 64, 548.
Novikov, D.N., Polikarpov, N.A., Deshevaye, A., et al., 2011. Acta Astronaut. 68, 1574.
Nuevo, M., Auger, G., Blanot, D., et al., 2008. Orig. Life Evol. Biosph. 38, 37.
Nuevo, M., Bredehöft, J.H., Meierhenrich, U.J., et al., 2010. Astrobiology 10, 245.
Nuevo, M., Materese, C.K., Sandford, S.A., 2014. Astrophys. J. 793, 125.

Nuevo, M., Cooper, G., Sandford, S.A., 2018. Nat. Commun. 9, 5276.

Nutman, A., Bennett, V., Friend, C., et al., 2016. Nature 537, 535.

Oba, Y., Takano, Y., Naraoka, H., et al., 2019. Nat. Commun. 10, 4413.

O'Brian, P.A., Houghton, J.A., 1982. Photochem. Photobiol. 35, 359.

Ohtomo, Y., Kakegawa, T., Ishida, A., et al., 2014. Nat. Geosci. 7, 25.

Olson, J.M., Pierson, B.K., 1986. Photosynth. Res. 9, 251.

Olson, S.L., Schwieterman, E.W., Reinhard, C.T., et al., 2018. Astrophys. J. 858, L14.

Oró, J., 1967. In: Fox, S.W. (Ed.), Origins of Prebiological Systems and of Their Molecular Matrices. Academic Press, New York, p. 137.

Osterbrock, D.E., Ferland, G.J., 2006. Astrophysics of Gas Nebulae and Active Galactic Nuclei. University science books.

Owen, J.E., Ercolano, B., Clarke, C.J., et al., 2010. MNRAS 401, 1415.

Owen, J.E., Clarke, C.J., Ercolano, B., 2012. MNRAS 422, 1880.

Owen, J.E., Wu, Y., 2013. Astrophys. J. 775, 105.

Owen, J.E., Wu, Y., 2017. Astrophys. J. 847, 29.

Owttrim, G.W., Coleman, J.R., 1989. J. Bacteriol. 171, 5713.

Papoular, R., Conard, J., Guillois, O., et al., 1996. A&A 315, 222.

Parker, E.T., Cleaves, H.J., Dworkin, J.P., et al., 2011. Proc. Natl. Acad. Sci. U. S. A. 108, 5526.

Pascucci, I., Gorti, U., Hollenbach, D., et al., 2006. Astrophys. J. 651, 1177.

Pearce, B.K.D., Pudritz, R.E., Semenov, D.A., et al., 2017. Proc. Natl. Acad. Sci. U. S. A. 114, 11327.

Pedersen, A., Gómez de Castro, A.I., 2011. Astrophys. J. 740, 77.

Peeples, M.S., Werk, J.K., Tumlinson, J., et al., 2014. Astrophys. J. 786, 54.

Pizzarello, S., Croning, J.R., 2000. Geochem. Cosmochim. Acta 64, 329.

Pizzarello, S., Schraderb, D., Monroea, A., et al., 2012. Proc. Natl. Acad. Sci. U. S. A. 109, 11949.

Planavsky, N.J., Reinhard, C.T., Wang, X., et al., 2014. Science 346, 635.

Planavsky, N.J., Slack, J.F., Cannon, W.F., et al., 2018. Chem. Geol. 483, 581.

Pollack, J.B., Hollenbach, D., Beckwith, S., et al., 1994. Astrophys. J. 421, 615.

Poppenhaeger, K., Robrade, J., Schmitt, J.H.M.M., 2010. A&A 515, A98.

Poppenhaeger, K., Schmitt, J.H.M.M., Wolk, S.J., 2013. Astrophys. J. 773, 62.

Prasad, S.S., Tarafdar, S.P., 1986. Astrophys. J. 267, 603.

Prochaska, J.X., Gawiser, E., Wolfe, A.M., et al., 2003. Astrophys. J. 595, L9.

Pudritz, R.E., Ouyed, R., Fendt, C., et al., 2007. Protostars and Planets V, p. 277.

Rai, R.K., Rastogi, S., 2010. MNRAS 401, 2722.

Ramirez, R.M., Kaltenegger, L., 2014. Astrophys. J. 797, L25.

Raymond, J.C., Blair, W.P., Long, K.S., 1997. Astrophys. J. 489, 314.

Reinhard, C.T., Planavsky, N.J., Olson, S.L., et al., 2016. Proc. Natl. Acad. Sci. U. S. A. 113, 8933.

Reinhard, C.T., Olson, S.L., Schwieterman, E.W., et al., 2017. Astrobiology 17, 287.

Ribas, I., Guinan, E.F., Güdel, M., et al., 2005. Astrophys. J. 622, 680.

Ribas, I., Bolmont, E., Selsis, F., et al., 2016. A&A 596, A111.

Ribas, I., Gregg, M.D., Boyajian, T.S., et al., 2017. A&A 603, A58.

Rice, W.K.M., Lodato, G., Armitage, P.J., 2005. MNRAS 364, L56.

Ripperda, B., Porth, O., Xia, C., et al., 2017. MNRAS 467, 3279.

Roberge, A., Feldman, P.D., Weinberg, A.J., et al., 2006. Nature 441, 724.

Roberge, A., Welsh, B.Y., Kamp, I., et al., 2014. Astrophys. J. 796, L11.

Robinson, T.D., Meadows, V.S., Crisp, D., et al., 2011. Astrobiology 11, 393.
Robinson, T.D., Ennico, K., Meadows, V.S., et al., 2014. Astrophys. J. 787, 171.
Rogers, L.A., Seager, S., 2010. Astrophys. J. 712, 974.
Romanova, M.M., Ustyugova, G.V., Koldoba, A.V., et al., 2004. Astrophys. J. 616, L151.
Romanova, M.M., Ustyugova, G.V., Koldoba, A.V., et al., 2012. MNRAS 421, 63.
Rosenbush, V., Kolokolova, L., Lazarian, A., et al., 2007. Icarus 186, 317.
Rosotti, G.P., Clarke, C.J., Manara, C.F., et al., 2017. MNRAS 468, 1361.
Rothschild, L.J., Mancinelli, R.L., 2001. Nature 409, 1092.
Rugheimer, S., Kaltenegger, L., Segura, A., et al., 2015. Astrophys. J. 809, 57.
Rybicki, G.B., Lightman, A.P., 2008. Radiative Processes in Astrophysics. John Wiley & Sons.
Salyk, C., Pontoppidan, K.M., Blake, G.A., et al., 2008. Astrophys. J. 676, L49.
Salyk, C., Blake, G.A., Boogert, A.C.A., et al., 2011. Astrophys. J. 743, 112.
Salz, M., Schneider, P.C., Fossati, L., et al., 2019. A&A 623, A57.
Sanchis-Ojeda, R., Rappaport, S., Winn, J.N., et al., 2014. Astrophys. J. 787, 47.
Sancho, L.G., de la Torre, R., Horneg, G., et al., 2007. Astrobiology 7, 3.
Saslow, W.C., Gaustad, J.E., 1969. Nature 221, 160.
Schindhelm, E., France, K., Herczeg, G.J., et al., 2012. Astrophys. J. 756, L23.
Schwarz, K.R., Bergin, E.A., Cleeves, L.I., et al., 2016. Astrophys. J. 823, 91.
Schwieterman, E.W., Robinson, T.D., Meadows, V.S., et al., 2015. Astrophys. J. 810, 57.
Schwieterman, E.W., Kiang, N.Y., Parenteau, M.N., et al., 2018. Astrobiology 18, 663.
Seager, S., 2014. Proc. Natl. Acad. Sci. U. S. A. 111, 12634.
Seinfeld, J.H., Pandis, S.N., 2016. Atmospheric Chemistry and Physics: From Air Pollution to Climate Change. John Wiley and Sons.
Selsis, F., Chazelas, B., Bordé, P., et al., 2007. Icarus 191, 453.
Shaikhislamov, I.F., Khodachenko, M.L., Lammer, H., et al., 2016. Astrophys. J. 832, 173.
Sicilia-Aguilar, A., Hartmann, L., Calvet, N., et al., 2006. Astrophys. J. 638, 897.
Sickafoose, A.A., Colwell, J.E., Horányi, M., et al., 2000. Phys. Rev. Lett. 84, 6034.
Singh, S.P., Hader, D.-P., Sinha, R.P., 2010. Ageing Res. Rev. 9, 79.
Sharma, M., Nath, B.B., Shchekinov, Y., 2011. Astrophys. J. 736, L27.
Shock, E.L., Schulte, M.D., 1990. Geochem. Cosmochim. Acta 54, 3159.
Shustov, B.M., Vibe, D.Z., 1995. ARep 39, 578.
Skinner, S.L., Schneider, P.C., Audard, M., et al., 2018. Astrophys. J. 855, 143.
Solhaug, K.A., Gauslaa, Y., Nybakken, L., et al., 2003. New Phytol. 158, 91.
Spitzer, L., 1978. Physical Processes in the Interstellar Medium. Wiley and Sons Inc.
Stecher, T.P., Donn, B., 1965. Astrophys. J. 142, 1681.
Tian, F., Toon, O.B., Pavlov, A.A., et al., 2005. Astrophys. J. 621, 1049.
Tian, F., Kasting, J.F., Liu, H.-L., et al., 2008. JGR (Planets) 113, E05008.
Tian, F., France, K., Linsky, J.L., et al., 2014. Earth Planet Sci. Lett. 385, 22.
Tielens, A.G.G.M., Allamandola, L.J., 1987. In: Morfill, G.E., Scholer, M. (Eds.), NATO ASIC Proc. 210: Physical Processes in Interstellar Clouds, p. 333.
Tielens, A.G.G.M., 2020. Molecular Astrophysics. Cambridge University Press.
Trilling, D.E., Lunine, J.I., Benz, W., 2002. A&A 394, 241.
van Zadelhoff, G.-J., Aikawa, Y., Hogerheijde, M.R., et al., 2003. A&A 397, 789.
Vangioni, E., Dvorkin, I., Olive, K.A., et al., 2018. MNRAS 477, 56.
Viallon, J., Lee, S., Moussay, P., et al., 2015. Atm. Mes. Tech. 8, 1245.
Vidal-Madjar, A., Lemoine, M., Ferlet, R., et al., 1998. A&A 338, 694.
Vidal-Madjar, A., Lecavelier des Etangs, A., Désert, J.-M., et al., 2003. Nature 422, 143.

Vidal-Madjar, A., Désert, J.-M., Lecavelier des Etangs, A., et al., 2004. Astrophys. J. 604, L69.

Vidal-Madjar, A., Huitson, C.M., Bourrier, V., et al., 2013. A&A 560, A54.

Vidotto, A.A., Jardine, M., Opher, M., et al., 2011. MNRAS 412, 351.

Vidotto, A.A., Bourrier, V., 2017. MNRAS 470, 4026.

Von Rekowski, B., Brandenburg, A., 2004. A&A 420, 17.

Walker, S.I., Bains, W., Cronin, L., et al., 2018. Astrobiology 18, 779.

Wallace, L., Barth, C.A., Pearce, J.B., et al., 1970. JGR 75, 3769.

Walsh, M.M., Lowe, D.R., 1985. Nature 314, 530.

Ward, W.R., 1997. Astrophys. J. 482, L211.

Watson, W.D., 1972. Astrophys. J. 176, 103.

Weaver, H.A., Feldman, P.D., Festou, M.C., et al., 1981. Icarus 47, 449.

Weingartner, J.C., Draine, B.T., 2001. ApJSS 134, 263.

West, A.A., Hawley, S.L., Bochanski, J.J., et al., 2008. AJ 135, 785.

Williams, J.P., Cieza, L.A., 2011. ARA&A 49, 67.

Winstanley, N., Nejad, L.A.M., 1996. Ap&SS 240, 13.

Wirström, E.S., Geppert, W.D., Hjalmarson, A., et al., 2011. A&A 533, A24.

Wood, B.E., Müller, H.-R., Zank, G.P., et al., 2005. Astrophys. J. 628, L143.

Wood, B.E., Müller, H.-R., Redfield, S., et al., 2014. Astrophys. J. 781, L33.

Woosley, S.E., Weaver, T.A., 1995. ApJSS 101, 181.

Wordsworth, R., Pierrehumbert, R., 2014. Astrophys. J. 785, L20.

Wright, N., 2006. PASP 118, 1711.

Wu, Y., Lithwick, Y., 2013. Astrophys. J. 772, 74.

Yang, H., Herczeg, G.J., Linsky, J.L., et al., 2012. Astrophys. J. 744, 121.

Yang, W., Marshak, A., Varnai, T., et al., 2018. Rem. Sens. 10, 254.

Yu, M., Willacy, K., Dodson-Robinson, S.E., et al., 2016. Astrophys. J. 822, 53.

Zahnle, K., Claire, M., Catling, D., 2006. Geobiology 4, 271.

Zhen, J., Chen, T., Tielens, A.G.G.M., 2018. Astrophys. J. 863, 128.

Superflares UV impact on the habitability of exoplanets[a]

Raissa Estrela[1,2], Adriana Valio[2]

[1]*Jet Propulsion Laboratory, California Institute of Technology, Pasadena, CA, United States;*
[2]*Center for Radioastronomy and Astrophysics Mackenzie, Mackenzie Presbyterian University, São Paulo, Brazil*

1. Introduction

To be considered potentially habitable, the main requirement for a planet is to be at a distance that allows it to have liquid water on its surface, also known as the Habitable Zone (HZ). However, many other factors can threaten the habitability of a planet and one of them is the magnetic activity of its host star. Living around an active star can be very dangerous. Even though the Sun is considered mildly active among other stars, our advanced civilization is frequently affected by some of the phenomena resulting from its activity, such as solar flares and coronal mass ejections. Moreover, solar flares release a hazardous amount of radiation, mostly X-rays and ultraviolet light (UV), that ionizes Earth's upper atmosphere, interfering with radio communications, and can also be potentially dangerous to unprotected astronauts outside of the International Space Station.

The Kepler and K2 missions, covering wavelength range of 430-890 nm, showed that solar-like stars that are younger than our Sun can produce more frequently and much stronger flares. Some of them are up to 10.000 times stronger than those released by the Sun and are denominated "superflares" (Maehara et al., 2015). These young rapidly rotating stars show high level of activity probably due to an increased dynamo action. Cooler stars like K and M dwarfs can also generate energetic flares, and in particular fully convective late M dwarfs are very active. This is the case of our solar neighbors Proxima-Centauri and TRAPPIST-1, two cool M dwarfs, which are known for hosting potentially habitable words. However, the frequent flares of the host stars could put the habitability of these worlds into question.

Superflares release significant amount of XUV, EUV, FUV, and UV radiation. Depending on the size of the flare, they can cause potential effects on the planetary atmosphere such as atmospheric loss and/or affect the chemical composition of the upper atmosphere. Moreover, protons accelerated in the flare produce odd nitrogen and odd hydrogen in the upper stratosphere and mesosphere that destroy ozone

---

[a] Fully documented templates are available in the elsarticle package on CTAN.

(Segura et al., 2010). Therefore, the energetic radiation from superflares could affect the origin and evolution of life on a planet orbiting in the HZ of the star.

The discovery of frequent energetic flares in young solar-like stars by the Kepler mission poses into question if such powerful flares were also common in the young Sun. As consequence, we start to wonder if they could have had an impact on the development of life in the primitive Earth. Nowadays, life forms in our planet are protected from most of the ionizing radiation of the solar flares by Earth's atmosphere, in particular due to the ozone layer. Despite being shielded by an ozone layer, today's civilization face several consequences if a superflare hits Earth. However, for the primitive Earth, such protection only started with the formation of the ozone layer 2500 Myr ago. Lingam and Loeb (2017) argue that powerful superflares can lead to extinction events, and their periodicity coincides with terrestrial fossil extinction record. On the other hand, Airapetian et al. (2016) suggested that superflares may have had a positive influence for the beginning of life on Earth. They showed that superflares associated with energetic particles can change the chemistry of the early Earth's atmosphere producing greenhouse gases responsible for warming our planet at a time when the Sun was not luminous enough. Therefore, young solar-type stars can be used to investigate the environment of the primitive Earth.

2. Kepler-96: a solar-type star with a Proterozoic age

Kepler-96 is a solar analogue star with an age of 2.3 Gyr which corresponds to the start of the Proterozoic Era on Earth. During that period, the oxygen level in Earth's atmosphere started to increase due to the photosynthesis of cyanobacteria living in the ocean. The accumulation of the oxygen allowed the formation of the ozone layer, responsible for absorbing most of the solar ultraviolet radiation arriving at Earth, in particular the most threatening radiation for life like UVC (100—280 nm) and UVB (280—315 nm), with only UVA (315—400 nm) reaching the surface. Even though planets can follow different evolutionary paths, a planet orbiting Kepler-96 could have had enough time to develop multicellular or single-celled life forms (such as cyanobacteria on Earth) 60 that helps the production of an ozone layer.

In addition, Kepler-96 is still a very active star, showing a clear rotational modulation in its lightcurve. Many flares can be observed in the lightcurve of the star, and also in the transit light curves of the planet Kepler-96 b that orbits this star every 16.23 days. Estrela and Valio (2018) analyzed three of these events and estimated the total energy from each flare by modeling the time profile as a Gaussian. The strongest flare, visible in the middle of the 48th transit (966.70 BJD—2.454.833 days), released a total of 1.8×10^{35} ergs, corresponding to the range of superflares (Maehara et al., 2015), is shown in Fig. 3.1.

This particular characteristic makes this star an interesting target to be used as a proxy to understand both: (1) the primitive Earth environment, assuming the Sun could also generate superflares at that age, and (2) a planet in the HZ with Archean conditions, considering that this planet had already enough time for life to evolve.

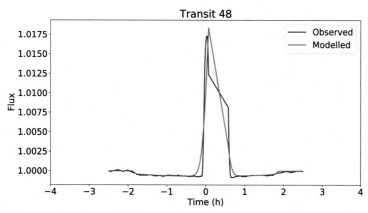

FIGURE 3.1

Strongest superflare observed on the 48th transit of Kepler-96. A Gaussian profile (pink) is used to model the time profile of the flare and to estimate its properties.

3. TRAPPIST-1 system

The TRAPPIST-1 system is one of the favorite candidates to host life. This system consists of seven planets orbiting an ultracool red dwarf star, and three of them (planets e, f, and g) are in the HZ (Demory et al., 2016).

These planets are considered to be "Earth-like" due to their size, radii of $0.918 \pm 0.039 R_{\oplus}$, $1.045 \pm 0.0389 R_{\oplus}$, and $1.127 \pm 0.041 R_{\oplus}$ for planets e, f, and g, respectively. Additionally, better estimates of their densities show that all of them are consistent with being mostly made of rock (Grimm et al., 2018).

M dwarfs stars are the most numerous ones in our galaxy, and they spend about 10^{10} years in the main sequence which gives enough time for complex forms of life to develop and evolve in the orbiting planets. In particular, TRAPPIST-1 has an age between 5.4 and 9.8 Gyr, indicating that the system is older than our Solar System. However, a study by Vida et al. (2017) using the data from the Kepler spacecraft in the K2 program found a frequent flaring rate during the 80 days of observations of TRAPPIST-1. A total of 42 flares were detected and the strongest eruption, shown in Fig. 3.2, emitted energy with at roughly 10^{33} ergs in white light, which is more energetic than the largest flare ever recorded from the Sun. These energetic events could threaten the habitability of the planets in the system as they orbit much closer to their host star (0.029, 0.037, and 0.0451 AU, respectively) than Earth.

4. Planetary atmosphere

Depending on the composition of the atmosphere, different UV wavelengths ranges can be absorbed. Shorter UV wavelengths (10−200 nm) are absorbed in the top of

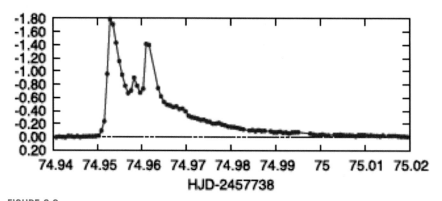

FIGURE 3.2

Strongest flare observed in TRAPPIST-1 with the K2 mission by Vida et al. (2017).

the atmosphere if the planet has strong absorbers like N_2, CO_2 or H_2O. While UV wavelengths within 200–300 nm (MUV) and 300–400 nm (NUV) can partially reach the surface of the planet depending whether the planet possess an ozone layer or not. In this context, the composition of the atmosphere is a key factor in determining the transmitted UV flux that reaches the surface of the planets.

Three types of atmospheres were considered by Estrela and Valio (2018) and Valio (2019) in the analyses of the transmitted UV flux to the surface of the hypothetical Earth orbiting Kepler-96 and the TRAPPIST-1 planets: an atmosphere similar to the Archean period on Earth and an atmosphere like the present-day Earth, with and without ozone. The present-day atmosphere is composed of 80% of N_2 and 20% of O_2, whereas the Archean atmosphere consists of 80% of N_2 and 20% CO_2.

5. Contribution of the flare to the UV flux

Estrela and Valio (2018) estimated the UV flux contribution of Kepler-96 flares using the MUV flux measured from the most intense solar flares, that was observed in 2003. The total thermal blackbody flux of Kepler-96 and the Sun are very similar, therefore Kepler-96 would increase the MUV flux proportionally to the Sun's flux. The solar flare of 2003, classified as an X17 GOES class, had a total energy of $E = 4 \times 10^{32}$ ergs, and a contribution of 23% to the VUV (0–200 nm) in the total solar irradiance (Woods et al., 2004). In addition, Kretzschmar et al. (2010) also analyzed the same event and claimed that the visible and the near UV (300–400 nm) in the flare spectrum are the dominant energy contributors, with the visible being responsible for 70% of the total irradiated energy. This means that the contribution to the total solar irradiation coming from the MUV (200–300 nm) is around 10%. Indeed, Woods et al. (2004) reported that the same solar flare increased by 12% the Mg II h and k emissions (279.58–279.70 nm), which is within the MUV range. Thus, using the increase of 12% in the MUV

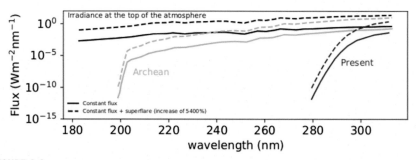

FIGURE 3.3

UV flux received by a hypothetical Earth-like planet orbiting Kepler-96 at the top of its atmosphere (black), or passing through an Archean (green) or present-day (purple) atmospheres. The contribution by the superflare is indicated by the dashed line.

flux, a superflare produced by Kepler-96 with $E = 1.8 \times 10^{35}$ ergs would increase by about 5400% the solar MUV flux (60% for $E = 2 \times 10^{33}$ ergs and 36% for $E = 1.2 \times 10^{33}$ ergs) of the young Sun. This increased UV flux is shown in Fig. 3.3 at the top of the atmosphere of the hypothetical Earth, and attenuated by Archean and present-day atmospheres.

In the case of TRAPPIST-1, a red dwarf star, the modeled spectra by O'Malley-James and Kaltenegger (2017) (black curves in Fig. 3.4) were used. The net UV flux that reaches the surface of the three TRAPPIST-1 planets in the HZ are shown for the different atmospheric types in Fig. 3.4.

In both Kepler-96 and TRAPPIST-1 planets, the Archean atmosphere absorbs only wavelengths smaller than 200 nm, allowing the harmful UVC irradiation to pass. On the other hand, the present-day atmosphere absorbs all the UVC irradiation due to the presence of an ozone layer. The increased UV flux incident at the top of the atmosphere of the TRAPPIST-1 planets is higher than those received by Kepler-96 because these planets orbit much closer to their host star.

6. Effective UV radiation on life

To determine the impact of the UV radiation on life we need to calculate the overall total photon energy in the UV band that falls in a unit area of a biological body, which is given by:

$$I_r = \int_0^T \int_{\lambda_1}^{\lambda_2} I(\lambda)d\lambda \, dt \tag{3.1}$$

where λ is the wavelength in the UV range, and T is the total duration of the UV exposition.

However, the response of a biological body varies as function of the wavelength. For example, the DNA molecules of living beings have higher response in the UVC

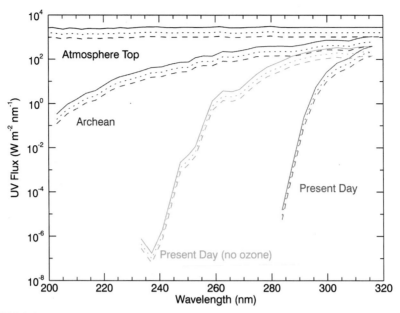

FIGURE 3.4

UV flux at the top of the atmosphere of planets TRAPPIST-1e (solid), TRAPPIST1f (*dotted*), TRAPPIST-1g (*dashed*) shown in black. The UV flux at the surface is transmitted by three atmospheric models: Archean (blue), present-day without ozone (green), and present-day atmosphere with ozone (magenta).

and UVB range. Therefore, it is necessary to weight the incident flux by the action spectrum, a function that express the biological response effectiveness at different wavelengths. Thus, we estimated the biologically effective irradiance (E_{eff}) by using the following expression:

$$E_{eff} = \int_{\lambda_1}^{\lambda_2} F_{inc}(\lambda)S(\lambda)d\lambda \qquad (3.2)$$

where F_{inc} is the total incident UV flux with the superflare contribution arriving at the planet surface, S is the action spectra and λ is the MUV wavelengths (200−300 nm).

Two bacteria have a very well-known action spectrum: *D. Radiodurans* and *Escherichia coli*. The first one is an extremophile, one of the most resistant organisms to radiation that can survive in extreme environments conditions like vacuum, dehydration, and acid. However, *E. Coli* has evolved to resist high pressures such as 2 GPa (Vanlint et al., 2011), and it has a genetic similarity with deep-sea bacterias (Horikoshi and Tsujii, 2012). The action spectra of the two bacteria are shown in Fig. 3.5. The UV dosage for 10% survival of *D. Radiodurans* is 553 J/m^2, which corresponds to a UV flux (255 nm) of 1.7 W/m^2 during 5 min (Ghosal et al., 2005), while for *E. Coli* it is 22 J/m^2.

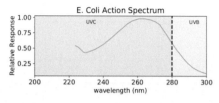
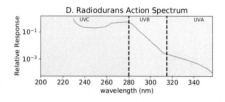

FIGURE 3.5

Action spectra, or biological response, for *E. Coli* (left) and *D. Radiodurans* (right) as shown in Estrela and Valio (2018).

7. Biological impacts on Kepler-96 and TRAPPIST-1 planets

7.1 On the surface

To analyze the impact on life due to the increase of the UV radiation by the super-flares, Estrela and Valio (2018) and Valio (2019) estimated the biologically effective irradiance, E_{eff} (Section 6), which gives information about the UV flux that falls into an unit area of the biological body.

Comparing the biological effective irradiance for these bacteria living in different atmospheric scenarios with the maximum UV tolerated by them, shown in Table 3.1, they found that the UV flux from the strongest superflare received by a biological body would only allow the presence of life on the surface of a hypothetical planet at one AU of Kepler-96 only if there was an atmosphere with ozone.

The same is true for TRAPPIST-1e, the planet in the HZ closer to the star. However, the two other planets, f and g, *D. Radiodurans* could survive even if there was no ozone in the atmosphere. Neither bacteria could survive in any planet with an Archean atmosphere due to the UVC radiation that reaches the surface.

Table 3.1 Biological effective irradiance, E_{eff} (J/m^2), due to the strongest superflare.

Planet	Bacteria	Archean	Present-day: no ozone	Present-day
Kepler-96 (1 AU)	E. coli	1.4×10^4	2.4×10^3	21.5
	D. radiodurans	8.0×10^3	1.3×10^3	7.5
TRAPPIST-1e	E. coli	9.0×10^3	962	8.2
	D. radiodurans	5.1×10^3	580	2.5
TRAPPIST-1f	E. coli	5.2×10^3	555	4.7
	D. radiodurans	2.9×10^3	334.4	1.5
TRAPPIST-1g	E. coli	3.3×10^3	353	3.0
	D. radiodurans	1.9×10^3	212	0.9

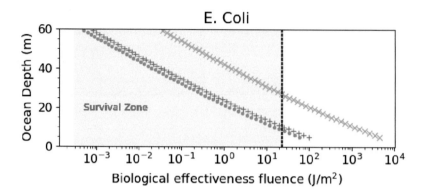

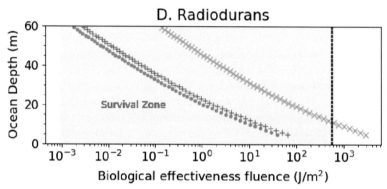

FIGURE 3.6

Top: Biologically effective irradiance for *E. Coli* (E_{eff}) with depth in an Archean ocean present in a planet in the HZ of Kepler-96. The values in Joules of the E_{eff} were obtained by multiplying the values in Watts with the time duration of the superflares, which are 7.1 min (flare A), 9.7 min (flare B) and 5.3 min (flare C). The E_{eff} computed with the contribution of each superflare is represented as follows: pink dot (flare C), purple cross (flare A), green x symbol (flare B). Bottom: The same for *D. Radiodurans*. The red vertical line represents the threshold for finding life determined by the maximum UV flux for 10% survival of *E. Coli* and of *D. Radiodurans*.

7.2 On the ocean

During the Archean era (3.9–2.5 Gyr ago), the ozone layer on Earth was being formed and therefore significant quantities of UVC and UVB radiation still reached the surface. This exposure to UV radiation could have imposed difficulties to the survival of microorganisms in the surface. Living in a deep ocean might have provided a safe refuge against the UV radiation. This could also be the case for Kepler-96 and the TRAPPIST-1 planets which receive much stronger UV radiation due to the superflares from the host star, the effects of the this radiation could be attenuated for microorganisms living in the ocean. For this reason, depending on the absorption of the UV radiation by the water, the aquatic environment is more likely to host life. The biological effective irradiance can also be calculated for an organism living in the ocean. In this case, we need to compute the UV irradiation at a certain ocean depth z and convolve it with the action spectrum. The UV spectral irradiance can be calculated using the following equation:

$$I(\lambda, z) = I_0(\lambda)e^{-K(\lambda), z} \tag{3.3}$$

where $I(\lambda,z)$ is the UV spectral irradiance at depth z, $I_0(\lambda)$ is the UV spectral irradiance with the superflare contribution passing through an Archean atmosphere and reaching the water surface and $K(\lambda)$ is the diffuse attenuation coefficient for water given by the sum of the absorption coefficient of water and the scattering coefficient.

Estrela and Valio (2018) assumed that the hypothetical Earth at one AU orbiting the star Kepler-96 has a calm and flat Archean ocean and found that *D. Radiodurans* and *E. Coli* would need to live at a depth of 12 and 28m, respectively. This result is shown in Fig. 3.6, where the threshold determined by the maximum UV flux for 10% survival of the bacteria is represented by a red vertical line.

7.3 Impact of consecutive superflares on the ozone layer

The work of Estrela and Valio (2018) and Valio (2019) show that the ozone layer is crucial to absorb most of the UV radiation of a single shot of irradiation from a superflare. However, the ozone column in the atmosphere can be affected by flares, especially if they are associated with energetic particles.

Previous works from Segura et al. (2010) analyzed the effects that the great flare observed on the active M-dwarf AD Leo could have on the composition of the atmosphere of an Earth-like planet orbiting at 1AU of this star. They showed that the recovery of the atmosphere depends whether the flare is associated with energetic particles. The ozone layer can recover in about 30 h if there are no energetic particle enhancement, otherwise the ozone layer could be depleted for about 30 years.

More recently, Tilley et al. (2019) investigated the impact of consecutive flares on the ozone layer of an Earth-like planet. They found that single lower-energy flares do not significantly impact the ozone column. Even for superflares with energies of 10^{34} erg separated by a time interval of 2 h, the impact on the ozone column is lower than the impact by a single proton event associated with a flare in this energy range. However, if lower-energy flares (3×10^{30} erg) associated with proton events occur in a frequency higher than once per year, they can rapidly erode the ozone column.

Acknowledgments

This research was carried out at JetPropulsion Laboratory, California Institute of Technology, under a contract with the National Aeronauticsand Space Administration (80NM0018D004). 2019 Allrights reserved.

References

Airapetian, V.S., Glocer, A., Gronoff, G., Hébrard, E., Danchi, W., 2016. Prebiotic chemistry and atmospheric warming of early Earth by an active young Sun. Nat. Geosci. 9, 452−455. https://doi.org/10.1038/ngeo2719.

Demory, B.-O., Queloz, D., Alibert, Y., Gillen, E., Gillon, M., 2016. Probing TRAPPIST-1-like systems with K2, 825, L25. https://doi.org/10.3847/2041-8205/825/2/L25.arXiv: 1606.08622.

Estrela, R., Valio, A., 2018. Superflare ultraviolet impact on Kepler-96 system: a Glimpse of habitability when the ozone layer first formed on Earth. Astrobiology 18, 1414−1424. https://doi.org/10.1089/ast.2017.1724.arXiv:1708.05400.

Ghosal, D., Omelchenko, M.V., Gaidamakova, E.K., Matrosova, V.Y., Vasilenko, A., Venkateswaran, A., Zhai, M., Kostandarithes, H.M., Brim, H., Makarova, K.S., et al., 2005. How radiation kills cells: survival of deinococcus radiodurans and shewanella oneidensis under oxidative stress. FEMS Microbiol. Rev. 29, 361−375.

Grimm, S.L., Demory, B.-O., Gillon, M., Dorn, C., Agol, E., Burdanov, A., Delrez, L., Sestovic, M., Triaud, A.H.M.J., Turbet, M., 2018. The nature of the TRAPPIST-1 exoplanets, 613, A68. https://doi.org/10.1051/0004-6361/201732233.arXiv:1802.01377.

Horikoshi, K., Tsujii, K., 2012. Extremophiles in Deep-Sea Environments. Springer Science & Business Media.

Kretzschmar, M., De Wit, T.D., Schmutz, W., Mekaoui, S., Hochedez, J.-F., Dewitte, S., 2010. The effect of flares on total solar irradiance. Nat. Phys. 6, 690.

Lingam, M., Loeb, A., 2017. Risks for Life on Habitable Planets from Superflares of Their Host Stars, vol. 848, p. 41. https://doi.org/10.3847/1538-4357/aa8e96.arXiv:1708.04241.

Maehara, H., Shibayama, T., Notsu, Y., Notsu, S., Honda, S., Nogami, D., Shibata, K., 2015. Statistical properties of superflares on solar-type stars based on 1-min cadence data. Earth Planets Space 67, 59. https://doi.org/10.1186/s40623-015-0217-z.arXiv:1504.00074.

O'Malley-James, J.T., Kaltenegger, L., 2017. UV Surface Habitability of the TRAPPIST-1 System, vol. 469, pp. L26−L30. https://doi.org/10.1093/mnrasl/slx047.arXiv: 1702.06936.

Segura, A., Walkowicz, L.M., Meadows, V., Kasting, J., Hawley, S., 2010. The effect of a strong stellar flare on the atmospheric chemistry of an Earth-like planet orbiting an M dwarf. Astrobiology 10, 751−771. https://doi.org/10.1089/ast.2009.0376.arXiv: 1006.0022.

Tilley, M.A., Segura, A., Meadows, V., Hawley, S., Davenport, J., 2019. Modeling repeated M dwarf flaring at an Earth-like planet in the habitable zone: atmospheric effects for an unmagnetized planet. Astrobiology 19, 64−86. https://doi.org/10.1089/ast.2017.1794.

Valio, E.R.C.L.G.A.,A., 2019. Accepted by the Proceedings of the International Astronomical Union.

Vanlint, D., Mitchell, R., Bailey, E., Meersman, F., McMillan, P.F., Michiels, C.W., Aertsen, A., 2011. Rapid acquisition of gigapascal-high-pressure resistance by *Escherichia coli*. mBio 2 e00130−10.

Vida, K., Kővári, Z., Pál, A., Oláh, K., Kriskovics, L., 2017. Frequent Flaring in the TRAPPIST-1 System—Unsuited for Life?, vol. 841, p. 124. https://doi.org/10.3847/1538-4357/aa6f05.arXiv:1703.10130.

Woods, T.N., Eparvier, F.G., Fontenla, J., Harder, J., Kopp, G., McClintock, W.E., Rottman, G., Smiley, B., Snow, M., 2004. Solar irradiance variability during the october 2003 solar storm period. Geophys. Res. Lett. 31.

Ultraviolet investigations of the interstellar medium from astrospheres to the local cavity

4

Jeffrey L. Linsky

JILA, University of Colorado and NIST, Boulder, CO, United States

1. Why is ultraviolet spectroscopy important for understanding the interstellar medium?

That most intrepid spacecraft *Voyager-1* is now directly sampling interstellar gas beyond the heliopause, where nonthermal solar wind ions interact with the inflowing interstellar hydrogen atoms. Now approaching a distance of 150 au, this spacecraft and its companion *Voyager-2* will not have the power or telemetry to communicate back measurements of the pristine interstellar gas beyond about 600 au that is not influenced by the solar wind. While a future interstellar probe mission could eventually measure the properties of interstellar gas beyond that distance, in the meantime our knowledge of the interstellar medium must be based on remote observations, primarily spectroscopic, and theory.

The extremely low density of interstellar gas, in the range $0.01-10^4$ cm^{-3}, means that collisional excitation rates are very small. As a result, all atoms, ions, and molecules will be in their ground states with few exceptions. Photoexcitation of most neutral and singly ionized atoms will, therefore, be from the ground state to an excited state with energies in excess of 4 eV, corresponding to ultraviolet wavelengths. The few exceptions are the resonant lines of low abundance species such as neutral Na and K and singly ionized Ca. The resonance lines of the abundant neutral atoms H, C, N, O, Mg, Si, and Fe and many of their first ions are located in the UV.

Stars provide bright back-illumination for studying the interstellar gas in absorption. Interstellar spectral lines are typically narrow as collisional broadening is negligible at the low densities in the ISM, and except for H, He, and O, interstellar absorption lines are optically thin or not very optically thick. Line broadening is, therefore, a combination of thermal velocities and the distribution of nonthermal flow velocities along the line of sight to the illuminating star. Narrow interstellar absorption lines are easily detected against the continuum or broad emission and absorption lines in stellar spectra. Ultraviolet spectroscopy with high spectral resolution is, therefore, the primary tool for studying the physical properties of interstellar gas.

Ultraviolet Astronomy and the Quest for the Origin of Life. https://doi.org/10.1016/B978-0-12-819170-5.00005-1

Table 4.1 Spectral regions, recent UV missions, and main species.

X-rays	<10 nm	Chandra, XMM-Newton	Fe IX-XXVI
EUV	10–91.2 nm	EUVE	He I, He II, Fe IX-XXIV
FUV	91.2–180 nm	HST, GALEX, IUE, FUSE	H I, C II-IV, N V, O VI, Si II-IV
NUV	180–320 nm	HST, GALEX, IUE	Mg II, Fe II
Ultraviolet C	100–280 nm		
Ultraviolet B	280–315 nm		
Ultraviolet A	315–400 nm		

Table 4.1 lists the usual wavelength ranges for the extreme-ultraviolet (EUV), far-ultraviolet (FUV), and near-ultraviolet (NUV) portions of the ultraviolet spectrum. Also listed are the main species located in these wavelength ranges. The wavelength intervals for Ultraviolet C, B, and A are commonly used for the study of radiation effects on organic materials including human skin.

2. Tools of the trade: satellites and instruments

Since ultraviolet light does not penetrate to the Earth's surface, observatories above the Earth's atmosphere are required. Table 4.2 is a nearly complete list of satellites and other instruments that have obtained UV spectra or broadband UV photometry. Beginning with the *Copernicus* satellite, there have been many instruments with sufficient sensitivity and spectral resolution to resolve interstellar absorption lines from background stellar features. The ability to detect and resolve weak interstellar absorption features has increased enormously since the early instruments because of simultaneous wide spectral coverage, reduced scattered light and detector noise, higher instrumental throughput, and increased telescope aperture (2.4m for *HST*). During its 18 year lifetime, the *International Ultraviolet Explorer (IUE)* pioneered observations of the ISM in the lines of sight to nearby stars, and the *Far-Ultraviolet Spectrograph Explorer (FUSE)* observed interstellar absorption lines in the 91–118 nm spectral region, which is instrumentally difficult because most optical surface coatings are poor reflectors in this spectral range. Fig. 4.1 shows a *FUSE* spectrum including interstellar absorption by C I, O VI, molecular hydrogen (H_2), and HD in the line of sight to the star HD 101190.

The main sources of interstellar absorption line spectra have been the three UV spectrographs on the *Hubble Space Telescope (HST)*: the High-Resolution Spectrograph (HRS), Space Telescope Imaging Spectrograph (STIS), and Cosmic Origins Spectrograph (COS). Their lifetimes and highest spectral resolutions are listed in Table 4.2. With a spectral resolution, $R = \lambda/\Delta\lambda = 80,000$, corresponding to a velocity resolution of 3.75 km/s with its echelle gratings, HRS could resolve interstellar absorption lines to measure velocities and line widths, $b = \sqrt{(\xi^2 + 2kT/m)}$, where T is

Table 4.2 Nonsolar UV observatories.

Observatory	Dates	Spectral region (nm)	$R = \lambda/\Delta\lambda$	Reference
Rockets	1945−present	NUV, FUV	Range	Morton (1967)
Copernicus (OAO3)	1972−1981	95−300	20,000	Rogerson et al. (1973)
Balloons (BUSS)	1976−1978	NUV (MgII, etc.)	27,000	Kondo et al. (1979)
IUE	1978−1996	115−330	12,000	Boggess et al. (1978)
IMAPS	1985−1996	93−115	150,000	Jenkins et al. (1996)
HUT	1990, 1995	83−185	400	Davidsen et al. (1992)
HST/GHRS	1990−1997	105−320	80,000	Brandt et al. (1994)
EUVE	1992−2001	7−76	200	Bowyer and Malina (1991)
ORFEUS	1993, 1996	40−128	3200	Hurwitz et al. (1998)
HST/STIS	1997−2004	115−320	114,000	Woodgate et al. (1998)
FUSE	1999−2007	90−119	20,000	Moos and Sonneborn (2006)
XMM-Newton/OM	1999−present	170−650	100	Mason et al. (2001)
SPEAR	2003	90−175	550	Edelstein et al. (2006)
GALEX	2003−2013	134−283	2 bands	Morrissey et al. (2005)
SWIFT/UVOT	2004−present	170−600	100	Mason et al. (2004)
HST/STIS	2009−present	115−320	114,000	Woodgate et al. (1998)
HST/COS	2009−present	115−320	18,000	Green et al. (2012)
ESCAPE	2025?	6.5−180	400	France et al. (2019)
WSO-UV/ VUVES,UVES	2025?	115−315	$\approx 50,000$	Shustov et al. (2018)
LUVOIR/ LUMOS	2030s?	100−1000	56,000	Concept Report (2019)
LUVOIR/ POLLUX	2030s?	100−400	120,000	Concept Report (2019)
HabEx/UVS	2030s?	115−320	60,000	Final Report (2019)

the temperature, m is the atomic mass, k is Boltzmann's constant, and ξ is the nonthermal broadening (often called turbulence) produced by the distribution of gas flows along the line of sight.

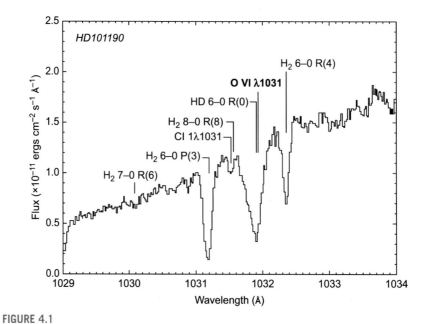

FIGURE 4.1

A *FUSE* spectrum showing interstellar absorption lines of C I, O VI, H₂, and HD in the line of sight to the O6 IV((f)) star HD101190 observed by Bowen et al. (2008)

STIS is now the prime UV instrument for interstellar spectroscopy because of the broad range of its powerful capabilities. With its highest resolution echelle gratings (E140H in the FUV and E230H in the NUV), it has achieved a resolution of $R = 114,000$, corresponding to a wavelength resolution of 2.63 km/s, with low detector noise and minimal scattered light. STIS also has lower resolution modes with higher throughput to achieve higher signal-to-noise on fainter targets. The moderate resolution echelle modes (E140M and E230M) have $R = 46,000$ corresponding to a velocity resolution of 6.5 km/s, which is adequate for many studies of the ISM. There are also lower resolution modes, a mode for imaging perpendicular to the wavelength direction (long slit spectroscopy), and a coronagraph mode. Placed on *HST* in 1999, STIS was operated until an electrical failure in 2004, and then refurbished by on-orbit replacement of its electronics in 2009. It is presently operating with its initial capabilities. Because it has a narrow slit that minimizes geocoronal emission, STIS has been the instrument of choice for observations of the important Lyman-α spectral line.

The third generation UV spectrograph on *HST*, the COS, was designed to obtain the highest possible throughput at moderate spectral resolution, $R \approx 18,000$, corresponding to a velocity resolution of 16.6 km/s. To achieve its enormous increase in throughput, the FUV mode has only one optical element that serves as the aberration corrector optic, disperser, and camera mirror. Compared to STIS with its six optical surfaces needed for high-resolution spectroscopy, the reduction to one reflecting surface increases the throughput of COS by more than an order of magnitude. Further increases in throughput result from the extremely low background in the detectors and a large entrance aperture rather than a narrow slit. However, COS's low spectral resolution and contamination of the Lyman-α line by geocoronal emission observed through the large aperture make STIS rather than COS the instrument of choice for interstellar spectroscopy.

Table 4.2 includes several telescopes with ultraviolet spectroscopic instruments that could fly in the 2020s or 2030s. The *World Space Observatory (WSO-UV)* is an approved Russian mission that could fly in the mid-2020s (Shustov et al., 2018). Two spectroscopic instruments, VUVES for the 115−170 nm range and UVES for the 170−315 nm range, are planned to have resolutions $R \approx 60,000$. The *Large Ultraviolet/Optical/Infrared (LUVOIR)* telescope is a major mission under study by NASA. As described in the Mission Concept Study Report (2019), *LUVOIR* is planned to include two UV spectroscopic instruments: the *LUVOIR* UV Multi Object Spectrograph (LUMOS) with resolution $R = 500 - 56,000$ in the 100−1000 nm range, and POLLUX, a UV spectropolarimeter operating in the 100−400 nm range with R up to 120,000. LUMOS would be a far more efficient successor to STIS. The *Habitable Exoplanet Observatory (HabEx)*, now under study by NASA with its primary objective to image exoplanets, will include a UV Spectrograph/Camera (UVS) to obtain $R = 60,000$ spectra in the 115−320 nm range as described in the HabEx Final Report (2019).

Low resolution spectrophotometry in the UV can study broad interstellar absorption features. For example, comparison of distant to nearby stars with the same spectral type reveals a broad absorption feature centered near 220 nm whose origin has been a subject of debate for many decades but could be due to organic dust in the ISM. Such measurements provide information on the particle size distribution of interstellar dust.

3. Challenges of the trade: interstellar absorption in the ultraviolet and extreme-ultraviolet

Absorption by interstellar gas and dust is both signal and obstacle. Narrow-band absorption lines produced by photoexcitation from the ground states of atoms, ions, and molecules reveal many of the physical properties of interstellar gas, whereas broadband absorption by dust and photoionization continua of abundant atoms like H attenuate the stellar emission flux that backlights interstellar absorption lines. Even though the mass ratio of gas to dust is typically about 100, dust absorption is important because it increases rapidly to the ultraviolet. Fitzpatrick and Massa (1986) and Fitzpatrick and Massa (1988) describe the broad interstellar feature centered at 217.5 nm seen in all directions in the Galaxy. They find that the central wavelength is the same within ± 1.7nm in all observations, but the FWHM varies from 36 nm through dark clouds to 60 nm in the diffuse ISM. The source of this absorption has been ascribed to small dust particles ($r \leq 50$Å) or a variant of fullerene (C_{60}). Recent studies of narrow absorption features in the UV and optical include, for example Schlafly et al. (2016) and Fitzpatrick et al. (2019).

The *Extreme-Ultraviolet Explorer (EUVE)* satellite detected the photoionization edges of neutral He ($\lambda < 50.4$ nm) and singly ionized He ($\lambda < 22.8$ nm) backlit by the bright EUV continua of hot white dwarfs. These spectra were the basis of the unexpected result that He is more highly ionized in the local ISM than H (Vallerga, 1996). Even the EUV spectrum of the nearest star Proxima-Centauri (d = 1.30 pc) (Craig et al., 1997) shows little flux at wavelengths $\lambda > 40$ nm because of interstellar H absorption. A proposed small satellite instrument called *ESCAPE (Extreme-ultraviolet Spectral Characterization for Atmospheric Physics and Evaluation)* (France et al., 2019) could extend interstellar absorption measurements with many times more sensitivity than *EUVE*.

The neutral H Lyman-α line (121.6 nm) plays a critical role in interstellar studies, because the abundance of other elements in interstellar gas are measured relative to H, and because Lyman-α is the only transition from the ground state of H that can be observed by *HST* in sightlines within the Galaxy. Observations of other Lyman series lines, e.g., Lyman-β (102.6 nm) and Lyman-γ (97.3 nm), and higher lines at shorter wavelengths, require either very large Doppler shifts possible only from extragalactic sources or satellite instruments with optical coatings suitable for observations at $\lambda < 115$ nm. The *FUSE* satellite has been the main source of moderate resolution Lyman series spectra. The *Copernicus* spacecraft also obtained such spectra, and for a few lines of sight the *IMAPS* instrument obtained high-resolution spectra of the Lyman series lines.

The analysis of Lyman-α line spectra provides a unique set of challenges. Fig. 4.2 shows the Lyman-α line observed from the star Capella (d = 13.12 pc) by the HRS instrument on *HST*. The intrinsic shape of the emission line emitted by the stellar chromosphere is roughly a Gaussian with broad wings, but interstellar neutral hydrogen in the line of sight to the star has an optical depth of about 10^6 that

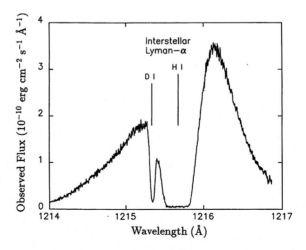

FIGURE 4.2

The GHRS high-resolution spectrum of Capella (G5 III + G0 III) showing the stellar Lyman-α emission line and interstellar absorption by hydrogen and deuterium Lyman-α.

Figure from Linsky, J.L., Brown, A., Gayley, K., et al., 1993. ApJ 402, 694.

produces a broad saturated absorption feature centered at the radial velocity of interstellar H in this direction. The intrinsic flux in the line emitted by the stellar chromosphere is, therefore, several times larger than the observed flux. The neutral hydrogen column density N(H I) in the star's line of sight can be modeled from the shape of the outer edges of the interstellar H absorption provided one has thermal and nonthermal broadening parameters obtained from the analysis of spectral lines of high-mass species such as Fe II (260 nm) and Mg II (280 nm) (Wood et al., 2005; Youngblood et al., 2016). As described in the next section, there is additional Lyman-α opacity produced in the heliosphere and astrospheres that broadens of the interstellar Lyman-α absorption. For log N(H I) greater than about 19, the far wings of the Lyman-α absorption can also provide reliable values of N(H I). Also shown in Fig. 4.2 is interstellar absorption by the deuterium Lyman-α line, which is optically thin or only moderately optically thick for lines of sight to nearby stars. The column density of D, N(D I), which can be obtained reliably from this line, provides a good estimate of N(H I) using the D/H abundance ratio D/H $= 15.6 \pm 0.4$ ppm (Linsky et al., 2006) in the Local Cavity.

4. Astrospheres: where the interstellar medium interacts with stellar winds

Measurements of neutral He atoms flowing into the solar system by the *IBEX* (Möbius et al., 2015) and *Ulysses* (Wood et al., 2015) spacecraft showed that the interstellar gas immediately surrounding the Sun has a temperature $T \approx 7,300$ K and a flow speed $v \approx 26$ km/s. This demonstrates that the Sun is moving though the local interstellar medium with a speed of about 26 km/s in the direction of the Scorpio-Centaurus Association. Along with the gas, interstellar dust is also flowing into the solar system with the same speed and direction (Strub et al., 2015). Neutral hydrogen atoms flowing into the solar system, however, have a slightly smaller speed, $v = 22 \pm 1$ km/s, higher temperature $T = 11,500 \pm 1,000$ K, and slightly different direction that the inflowing He (Lallement et al., 2005). The different properties of H and He result from charge exchange interactions between outflowing solar wind protons and the inflowing interstellar neutral H atoms. Since the charge exchange cross-section for protons with neutral He atoms is negligible, the incoming He atoms likely provide a reliable measurement of the gas properties in the local ISM.

The 400–800 km/s solar wind speed is supersonic in the million degree solar corona, and the 26 km/s interstellar gas speed is supersonic in the 7300 K local ISM. The interaction of these two supersonic flows is characterized by two shocks and an intervening subsonic region. Fig. 4.3 is an example of this interaction in the outer heliosphere computed with a kinetic plasma code (e.g., Müller et al., 2008). The inner shock, called the termination shock, occurs where the solar wind decelerates to a subsonic flow. *Voyager-1* crossed the termination shock at 94 au and

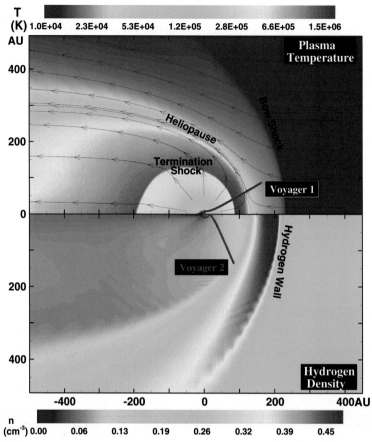

FIGURE 4.3

Plasma model of the heliosphere computed by Hans-Reinhardt Muller. **Top** Temperatures from the Sun (0 in the x axis) through the termination shock, heliosheath, heliopause, and beyond into the local ISM. The interstellar flow approaches the heliosphere from the right. There is a bow shock or bow wave at about 300 au in this model. **Bottom** Neutral hydrogen densities peak in the hydrogen wall at about 200 au upstream in this model.

Figure courtesy of H.-R. Muller.

Voyager-2 crossed it at 85 au. Both spacecraft then entered the heliosheath where the solar wind plasma is heated by the termination shock and hot pickup ions and accelerated ions are produced. The two spacecraft then headed outward to the heliopause at distances 121.6 au and 119.0 au, respectively, where both *Voyagers* measured ionized interstellar gas flowing past the Sun (Burlaga et al., 2019).

What they sampled was not pristine interstellar gas because charge exchange reactions between outflowing solar wind protons and the inflowing neutral H atoms. In these reactions solar wind protons become H atoms that are heated, slowed down, and become denser in the "hydrogen wall." As shown in Fig. 4.3, the neutral H density enhancement is only about a factor of three, but N(H I) in the hydrogen wall and the decreased flow speed by a few km/s relative to the interstellar gas inflow are sufficient to produce an observable absorption feature in the Lyman-α line that is redshifted relative to the interstellar gas absorption. In the heliosphere models of Zank et al. (2013), the peak neutral H density in the hydrogen wall occurs near 300 au and the bow shock or bow wave where interstellar gas slows down occurs near 500 au. Unfortunately, the *Voyagers* will not survive long enough to sample these features in the heliosphere. Beyond about 600 au, the gas in the local ISM (LISM) should not be affected by the solar wind and thus be pristine. Hydrogen wall absorption in the heliosphere was detected in the Lyman-α spectra for eight lines of sight and hydrogen wall absorption was detected from 13 astrospheres (stellar analogs of the heliosphere) by Wood et al. (2005). Several more heliosphere and astrosphere hydrogen walls have been detected since then.

The well-studied heliosphere provides the role model for the study of astrospheres in which the same physical processes should occur as in the heliosphere. The locations of the termination shock, astropause, and hydrogen wall together with the hydrogen column density and decelerated velocity in the hydrogen wall all depend upon the mass flux and speed of the stellar wind and the velocity of the star through the ISM. Since the H atoms in the hydrogen wall are slowed down relative to the interstellar gas, distant observers looking at an astrosphere will see the hydrogen wall gas blue shifted relative to the main interstellar absorption. This is illustrated by a comparison of the Lyman-α spectra of Proxima-Centauri (α Cen C) to that of α Cen B, which are only one degree apart in the sky and likely sample the same ISM gas flow. Fig. 4.4 shows that the long wavelength (red) side of the interstellar absorption is identical in both stars, but the long wavelength (blue) side is different for the two stars. Thus the hydrogen wall absorption in the two astrospheres must be different.

To proceed further requires the comparison of such spectra with theoretical astrosphere models. Wood et al. (2005b) computed plasma astrosphere models for these two stars for which the main difference with solar atmosphere models is the strength of the stellar wind measured by the mass-loss rate. Fig. 4.4 shows that with increasing mass-loss rate, the hydrogen wall absorption in the astrosphere adds additional absorption to the short wavelength edge of the interstellar H absorption. The location of the blue edge of the interstellar absorption is, therefore, a stellar mass-loss meter. This figure measures the combined mass-loss rates of the α Cen AB binary to be twice that of the Sun as α Cen A and B are both located inside a common astrosphere. The upper limit to the mass loss rate for Proxima-Centauri is one-tenth that of the Sun or $2 \times 10^{-15} M_\odot$ per year (Wood et al., 2005b). These results were the first observationally based mass-loss rates for solar-type dwarf stars other than the Sun. Fig. 4.5 shows Lyman-α spectra of other stars for which mass-loss rates as large as 60 times solar were measured by Wood et al. (2005b). The most recent compilation of stellar mass-loss rates using this technique by Wood et al. (2014)

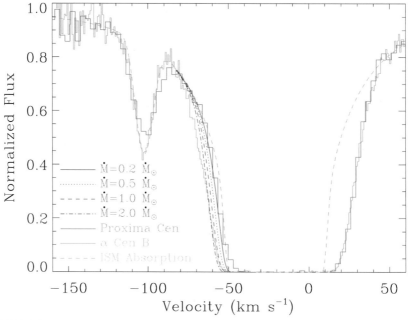

FIGURE 4.4

Comparison of the cores of the Lyman-α lines of α Cen B (blue) and Proxima-Centauri (red). The interstellar Lyman-α absorption is estimated from the shapes of the interstellar D I Lyman-α, Mg II, and Fe II lies (shown in green). On the red side (positive velocities) of the interstellar absorption is additional absorption from the heliospheric hydrogen wall (seen in the spectra of both stars). On the blue side (negative velocities) of the interstellar absorption Proxima-Centauri shows no additional absorption, but α Cen B shows additional absorption from the hydrogen wall of the combined α Cen AB astrosphere. Dashed lines show plasma models for different stellar mass-loss rates with twice that of the Sun a best fit. The interstellar D I Lyman-α absorption is not fully resolved in the Proxima-Centauri spectrum due to a low resolution spectrum of this faint star.

Figure from Wood, B.E., Linsky, J.L., Müller, H.R., Zank, G.P., 2001. ApJl 547, L49.

includes 11 stars showing a trend of mass-loss rates compared to X-ray surface flux. Unfortunately there are only a few mass-loss rate measurements for very active stars and M dwarfs.

Mass loss from stars in the form of stellar winds has been studied in hot stars using P-Cygni shape emission line profiles and free-free radio emission and in cool giants and supergiants by modeling blue-shifted absorption line features. These techniques do not work for F—M main-sequence stars that have very low mass-loss rates if the Sun is a useful guide ($\dot{M}_{\odot} = 2 \times 10^{-14} M_{\odot} \text{yr}^{-1}$). However, essentially all exoplanets presently discovered are companions of main-sequence stars with spectral types F—M rather than hot O-type stars or G—M giants and supergiants. The stellar winds that impact the outer atmospheres of exoplanets can be

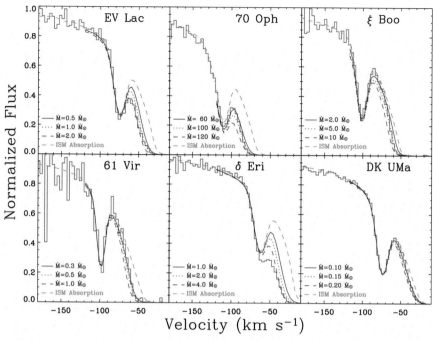

FIGURE 4.5

Lyman-α line profiles of six stars observed by the HRS and STIS instruments. Shown are the short wavelength portions of the line including the blue side of the interstellar hydrogen absorption and the interstellar deuterium absorption line. Dashed lines indicate additional absorption due to astrospheric hydrogen for different assumed mass-loss rates.

Figure from Wood, B.E., Müller, H.R., Zank, G.P., Linsky, J.L., Redfield, S, 2005b. ApJL 628, L143.

measured only using the astrospheres technique, estimated from theoretical models, or scaled from the mass-loss rates measured or estimated from the few stars that have measured winds.

As best we can tell, habitable exoplanets must have atmospheres in order to retain surface liquid water and provide a shield from stellar ultraviolet B and C radiation that destroys complex organic molecules such as DNA. Rocky Earths and super-Earths can lose their initial atmospheres early in their evolution from the very strong EUV radiation of their young host star, but the retention of an Earth-like secondary atmosphere likely needed for habitability depends in large part on the wind strength of more mature host stars. Magnetic stellar winds can capture protons and other ions in the outer atmospheres of planets (Linsky, 2019). For example, Mars likely lost its atmosphere through dissociative recombination reactions that create hot O, O^+, and CO_2^+ ions that were captured by the solar wind (Dong et al., 2018; Jakosky et al., 2018). For exoplanets that have orbits close to their host stars, e.g., V374 Peg b (Vidotto et al., 2011) and Proxima b (Garraffo et al.,

2016), very strong winds can strip the exoplanet's atmosphere in a short time. Strong planetary magnetic fields can reduce but not prevent mass loss as stellar wind ions can penetrate deep into a planet's atmosphere near the magnetic poles. Thus host star winds, measured by the astrosphere technique, play a critical role in determining whether an exoplanet could be habitable.

The study of astrospheres demonstrates the importance of measuring the kinetic properties of the local interstellar medium. A key parameter for mass-loss rate analysis is the direction and speed of the partially ionized interstellar gas that the star encounters. This information comes from models of the warm clouds in the local ISM (LISM).

5. Warm clouds in the local interstellar medium

From an analysis of interstellar absorption lines toward stars within 100 pc, Crutcher (1982) proposed that the LISM consists of coherently moving gas inflowing from the center of the Scorpio-Centaurus Association. Subsequently, Lallement and Bertin (1992) showed that toward stars in the direction opposite to the Galactic Center the interstellar gas flow has a different velocity vector that they called the anti-Center (AC) cloud. Later observations based on high-resolution *HST* spectra of interstellar Mg II, Fe II, and Lyman-α lines determined the shape and properties of the AC cloud, which is now called the Local Interstellar Cloud (LIC), and the cloud toward the Galactic Center, which is now called the G cloud.

The technique for identifying interstellar clouds and obtaining their shapes and physical properties is outlined in Fig. 4.6. Useful observations of interstellar spectral

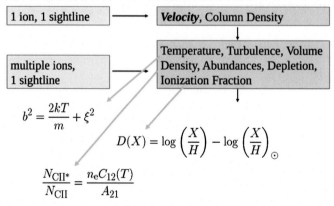

FIGURE 4.6

Observational diagnostics—the interstellar gas properties that can be obtained from high-resolution spectra of one ion or multiple ions in a single sightline showing the equations used to infer temperature, turbulence, element depletion, and electron density.

Figure courtesy of S. Redfield.

lines require sufficient spectral resolution, ideally better than 4 km/s, to resolve the line width and depth and an absolute velocity scale better than about 2 km/s to reliably intercompare different spectra. The high-resolution STIS echelle modes meet these stringent requirements. With such data, the analysis of one spectral line observed in one line of sight provides two parameters: the radial velocity of the interstellar gas, and the column density of the atom or ion. Since even for lines of sight shorter than 5 pc there is often structure in the absorption line signifying more than one velocity component, high signal-to-noise spectra of high-mass ions are needed to resolve the velocity structure. Fig. 4.7 shows an example of the interstellar velocity structure in the Mg II, Fe II, and D I Lyman-α lines seen in the spectrum of the star HD 9826 (d = 13.5 pc) (Edelman et al., 2019).

Spectra of several atoms or ions with different atomic weights in the same line of sight provides additional information. Since line widths are the sum of thermal motions in the gas, which are inversely proportional to atomic mass (m), and turbulent flows, which are independent of mass, the comparison of line widths of the heavy ions Mg II ($m = 12$) and Fe II ($m = 26$) with the light atom D I ($m = 2$) provides both the gas temperature and the turbulence in the line of sight. There are several approaches to measuring the electron density in a velocity component. For example, the ratio of the column density of a forbidden line to that of a permitted line in the same multiplet of an ion such as C II, or the column density ratio from different ionization stages of the same element such as Mg II relative to Mg I. Gry and Jenkins (2014) and Frisch et al. (2011) describe these methods in detail. The depletion of an element in interstellar gas relative to its cosmic abundance (usually assumed to be the solar abundance relative to hydrogen) provides a measure of the fraction of that element captured on grains and therefore removed from the interstellar gas.

Observations of interstellar gas properties in many lines of sight across the sky provided the data for identifying individual clouds and their shapes. Redfield and Linsky (2008) analyzed the radial velocities of 270 velocity components observed in the lines of sight to 157 stars. They first asked whether a single velocity vector could explain the radial velocities of all of these components. The least squares fit to the data showed that 45% of the components have radial velocity differences from the mean that are at least 10 times the velocity measurement errors. Clearly a single vector is a poor fit to the data. They then asked whether the radial velocities toward stars distributed over a wide angle can be fit by a single velocity vector. By systematically removing the most discrepant radial velocities one by one, they found that 79 velocity components centered in the anti-Galactic Center direction are well fit by a single velocity vector within measurement errors. They had found the outer boundary of the LIC, as shown in Fig. 4.8. They then removed the LIC velocity components from the list and asked whether there is another velocity vector consistent with the radial velocities of many components following the same procedure. Indeed there is a second vector centered in the Galactic Center region which they called the G cloud. Further searches identified a total of 15 clouds within 15 pc of the Sun. The closest star having an interstellar velocity component consistent with the velocity vector sets an upper limit to the distance of the cloud. There are likely other

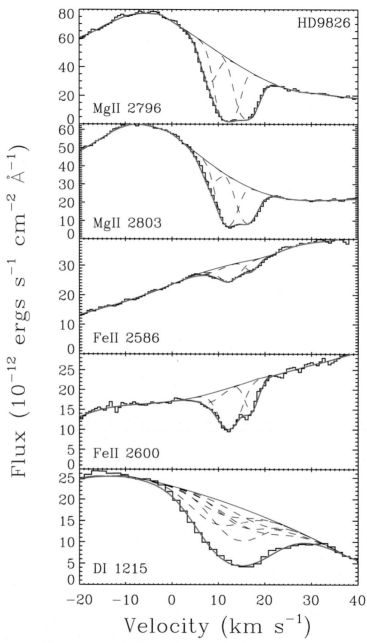

FIGURE 4.7

An example of a three velocity component fit to the star HD 9826 (F9 V). The separation of the three components is cleaner in the high-mass Fe II and Mg II spectral lines where thermal broadening is unimportant, but the three components are hard to distinguish in the D I Lyman-α line for which thermal broadening dominates.

Figure from Edelman, E., Redfield, S., Linsky, J.L., Wood, B.E., Müller, H., 2019. ApJ 880, 117.

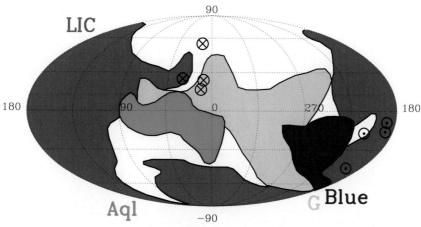

FIGURE 4.8

Morphologies of the four partially ionized VLISM clouds that are closest to the outer heliosphere. They are the LIC (red), which lies in front of ε Eri (3.2 pc), the G cloud (brown), which lies in front of α Cen (1.32 pc), the Blue cloud (dark blue), which lies in front of Sirius (2.64 pc), and the Aql cloud (green), which lies in front of 61 Cyg (3.5 pc). The plot is in Galactic coordinates with the Galactic Center direction in the center. The upwind direction of the LIC velocity vector is indicated by the circled-cross symbol near $l = 15°$ and $b = +20°$, and the upwind directions of the other clouds have similar marks. The downwind directions are indicated by the circled-dot symbols.

Figure from Redfield, S., Linsky, J.L., 2008. ApJ 673, 283.

members of the cluster of local interstellar clouds (CLIC) that are too small, too distant, or have too few data points for this technique to work. Fig. 4.8 shows the boundaries in Galactic coordinates of the four nearest clouds: the LIC, the G cloud seen in front of α Cen (1.32 pc), the Blue cloud seen in front of Sirius (2.64 pc), and the Aql cloud seen in front of ε Eri (3.5 pc).

Determining the boundaries of LISM clouds is an imperfect task. The morphology of the LIC is determined from 62 lines of sight with radial velocities consistent with the LIC kinematic vector. Linsky et al. (2019) determined the distance from the Sun through the LIC from the measured N(H I) along each sightline and the neutral hydrogen number density $n_{\mathrm{H\,I}} = 0.20$ cm^{-3} (Slavin and Frisch, 2008). Fitting these distances with nine spherical harmonics resulted in the three-dimensional model shown in Fig. 4.9. Distances for eight of the 62 lines of sight are either outside or inside of the spherical harmonic fit by $> 3\sigma$ suggesting that the assumption of homogeneous properties in the LIC may not be valid. Computing three-dimensional models for the G and other clouds is a challenge because there are fewer lines of sight through these clouds and the distances to the beginning and end through each cloud are not known. The Galactic coordinates for the lines of sight with radial velocities consistent with each cloud's kinematic vector provide constraints on the cloud boundaries. Redfield and Linsky (2008) drew approximate

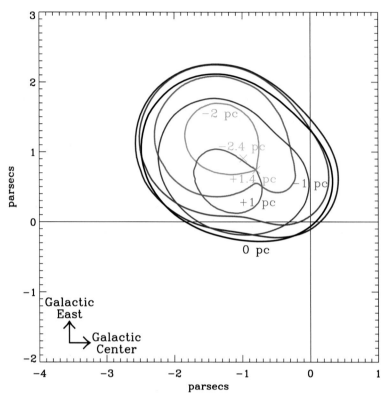

FIGURE 4.9

Contour map of LIC as viewed from North Galactic Pole. The Sun is located at the origin (0,0). Red contours are cuts below the LIC center parallel to the Galactic plane, and the blue contours are cuts above. The two x symbols indicate the locations where the edge of the LIC is furthest above and below the plane of the figure.

Figure from Linsky, J.L., Redfield, S., Tilipman, D., 2019. ApJ, 886, 41.

boundaries outside of these coordinates but inside of coordinates for lines of sight that did not show radial velocities consistent with the cloud's kinematic vector. This procedure produces good but approximate boundaries for clouds with many lines of sight (21 for G, 10 for Blue, and 9 for Aql). Four lines of sight were considered the minimum for determining a kinematic vector. Fig. 4.8 shows the morphologies of the four clouds that are likely to be closest to the Sun. Work to determine three-dimensional morphologies for several of the clouds is underway using additional lines of sight that are becoming available.

An important question is which model is more realistic: (1) the CLIC consists of discrete clouds each with its own velocity vector, or (2) the local gas is one entity with gradients in its velocity structure. One might assume the latter to be the case since densities in the local ISM are very low and the time scale for collisions is very long. On the other hand, the gas may be organized into individual velocity structures by magnetic fields or photoionization from one or several directions. Gry and Jenkins (2014) developed a model in which velocity gradients within the local ISM can explain the same observed radial velocities that Redfield and Linsky (2008) used to develop the 15-cloud model. A good test of the merits of the two models was provided by a new data set (Malamut et al., 2014) consisting of 34 lines of sight that were observed by an *HST* SNAP program in which the telescope schedulers selected stars from a large list to fit the availability of single orbits between large observing programs. All of the observed radial velocities obtained in this SNAP program are consistent within observational errors of the predictions of the cloud vectors in the 15-cloud model. Redfield and Linsky (2015) then compared the radial velocities of the SNAP stars predicted by the two models concluding that the 15-cloud models provides a better fit. The story is not yet finished, however, as observations along new lines of sight may lead to more complex models.

Another important question is whether the Sun is located inside of the LIC, at the outer edge of the LIC, or just outside of the LIC. Since the LIC covers about 45% of the sky, each of these options is possible. Additional evidence cited below probably resolves the question.

In Fig. 4.8 the clouds are represented in two dimensions, but the clouds must be three dimensional. Linsky, Redfield and Tilipman (2019) computed a three-dimensional model of the LIC using neutral H column densities N(H I) for 62 velocity components assuming that the distances through the LIC to stars are $d = N(H\ I)/n(H\ I)$ with the neutral hydrogen number density equal to $0.20\ \text{cm}^{-3}$ (Slavin and Frisch, 2008). The LIC model shown in Fig. 4.9 extends 2−3 pc with the Sun at or just beyond its edge. The speed and direction of He atoms flowing into the solar system can help resolve the question of the location of the heliosphere relative to the LIC. The He inflow speed is about 2 km/s faster than predicted by the LIC model with a direction 3 degrees off in ecliptic longitude and 2 degrees off in ecliptic latitude (Linsky et al., 2019). This strongly suggests that the Sun is located at the edge of the LIC where the flow is slightly different from the mean flow of the LIC.

The density, temperature, ionization, and other properties of the LIC are summarized in a model developed by Slavin and Frisch (2008) and Frisch et al. (2011). The model parameters are $T = 6,680 \pm 1,490$ K, n(H I) $= 0.19-0.20\ \text{cm}^{-3}$, and thermal pressure $P_{th}/k = 3,180^{+1,850}_{-1,130}$ Kcm^{-3}. The electron density is in the range $n(e) = 0.05-0.08\ \text{cm}^{-3}$, indicating that hydrogen in the LIC is about 30% ionized. These parameters are consistent with the warm component of the interstellar medium originally proposed by Field et al. (1969). There are a few measurements of electron densities for other clouds by Redfield and Falcon (2008) and Gry and Jenkins (2017), but the detection of many astrospheres with hydrogen walls leads me to assume that the other warm clouds in the local ISM are also partially ionized.

Linsky et al. (2019) computed the distribution of hydrogen column densities in all directions from the geometric center of the LIC using the three dimensional model. Fig. 4.10 shows the distances from the center of the LIC to its edge computed

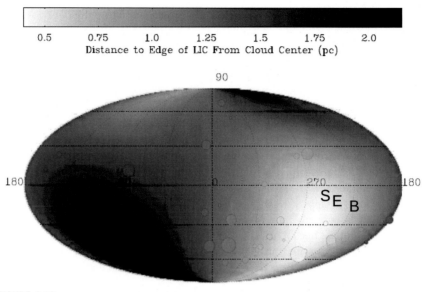

0.5   0.75   1.0   1.25   1.5   1.75   2.0
Distance to Edge of LIC From Cloud Center (pc)

FIGURE 4.10

Distances to the edge of the LIC from its geometric center computed from N(H I) along lines of sight to 62 nearby stars. The symbols are S for the Galactic coordinates of the hot white dwarf Sirius B, E for ε CMa, and B for β CMa. Faint circles indicate the locations of stars near the hydrogen hole.

Figure from Linsky, J.L., Redfield, S., Tilipman, D., 2019. ApJ, 886, 41.

from N(H I) divided by n(H I) = 0.20 cm^{-1}. There is a region of very low distance to the edge south of the Galactic equator between Galactic longitudes 225° and 280°, which they called the "hydrogen hole." Within the hydrogen hole are three important sources of EUV radiation (Vallerga, 1998): ε CMa, the brightest EUV source detected in the *EUVE* all sky survey, β CMa, the second brightest EUV source, and Sirius B a nearby hot white dwarf. The straight forward explanation for the hydrogen hole is that the EUV radiation from ε CMa and the other stars is photoionizing neutral hydrogen in the direction to these stars. Thus photoionization is playing an important role in shaping the clouds in the LIC. Fig. 4.11 shows the positions of the four nearest clouds in the CLIC relative to ε CMa and other stars. What role does photoionization play in the understanding the Local Cavity?

6. The local cavity

The Local Cavity is an irregularly shaped region of low density plasma extending 100–200 pc from the Sun and surrounded by higher density cool gas identified by interstellar absorption in lines of Na I and Ca II. A recent morphology of the

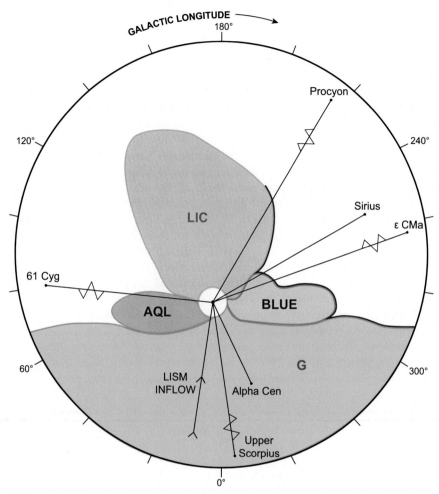

FIGURE 4.11

The local ISM region within 3 pc of the Sun as viewed from the North Galactic Pole showing the location of the four partially ionized clouds that are in contact with the outer heliosphere. Not shown are other clouds lying outside of the four clouds. Shown are the Sun (point), an exaggerated representation of the heliopause (circle around the Sun) and the LIC, G, Aql, and Blue clouds. Lines of sight projected on to the Galactic equator are shown for five stars. Red shading shows the Strömgren shells produced by EUV radiation from ε CMa. Also shown are the direction of inflowing interstellar gas as seen from the Sun and the direction to the Upper Scorpius region of the Scorpius-Centauri Association where the most recent supernovae likely occurred.

Figure from Linsky, J.L., Redfield, S., Tilipman, D., 2019. ApJ, 886, 41.

Local Cavity is presented in Capitanio et al. (2017). The Local Cavity contains the CLIC and perhaps other warm clouds that have not yet been detected, at least one cold cloud, the Leo Cold Cloud (Peek et al., 2011), and completely ionized hydrogen. The Local Cavity was originally called the Local Bubble, because it was assumed to consist of X-ray emitting hot plasma that was ionized and evacuated by supernova explosions. The basis for this designation is that supernovae occurred as recently as 1.5–3.2 Gyr ago (Wallner et al., 2016) in the Scorpio-Centaurus Association now 140 pc away. Shock waves from one or more supernova explosions heated the surrounding gas to temperatures near 10^6 K with radiative and adiabatic cooling thereafter. The total ionization of H, diffuse soft X-ray emission from the hot gas and partial ionization of the warm clouds by the X-radiation fit this scenario. It was generally accepted that the hot gas in the Local Bubble model is an example of the hot component of the ISM in the theoretical models of Wolfire et al. (1995).

This picture has been challenged by Welsh and Shelton (2009) and others for a number of reasons. One reason is that most of the soft X-ray emission is not from hot interstellar gas (except in the Galactic polar region), but rather from charge exchange reactions between solar wind protons and neutral H in the interplanetary medium (e.g., Koutroumpa, 2012). Another reason is that there must be an interface (either conductive or turbulent) between hot gas and the warm CLIC gas that should be detected by absorption from such ions as O VI and N V, but absorption lines from these ions have not been detected except from hot white dwarfs. Finally, searches for emission in Fe IX 17.1 nm and other emission lines predicted for the hot component gas were not detected by the *EURD* and *CHIPS* satellites (Edelstein et al., 2001; Hurwitz et al., 2005). While there are plausible explanations for these observational discrepancies, one should consider other models of the Local Cavity consisting of low density gas that is much cooler than 10^6 K in which H is fully ionized.

One class of alternative models is a nonequilibrium recombining plasma following supernova explosions (e.g., Lyu and Bruhweiler, 1996; Breitschwerdt and Schmutzler, 1999). Such models can avoid the need for X-ray emission from hot plasma and intermediate temperature ions like O VI and C IV at the interfaces between warm clouds and the Local Cavity plasma. The elapsed time since recent supernova plays an important role in such models.

A third type of model is a photoionized warm plasma produced by the EUV radiation of stars in the Local Cavity. Strömgren (1939) showed that the EUV radiation from a hot star or white dwarf would be surrounded by fully ionized H out to the distance where recombinations exceed H ionizations. This volume of ionized hydrogen is called a Strömgren sphere or an H II region and the distance to its edge is called the Strömgren radius. The outer region will have a thickness determined by the surrounding neutral hydrogen density and the recombination cross-section. For $n(\text{H I}) = 0.2\,\text{cm}^{-3}$, as is the case for the LIC, the thickness of a Strömgren shell is about 0.2 pc. This very simple picture neglects overlapping Strömgren spheres, dust absorption, He ionization, motions of the hot star through its Strömgren sphere, local density variations, and stellar variability, but the basic scenario is useful for understanding the Local Cavity.

Could the LIC and other warm clouds in the CLIC be inside the Strömgren sphere of the brightest EUV source ε CMa? The radius of a simple Strömgren sphere is

$$R^3 = 3(dN_i / dt)/(4\pi\alpha n_i n_e).$$

Here dN_i/dt is the number of ionizing photons per second, α is the hydrogen recombination factor 4×10^{-13}, and n_i and n_e are the number densities of protons and electrons in the sphere. Based on the measurement of the EUV flux from ε CMa and the assumption that $N(H\ I) = 9 \times 10^{17}$ cm^2 in the line of sight to the star, Vallerga (1998) estimated the intrinsic stellar flux to be $dN_i/dt = 2.7 \times 10^{46}$ EUV photons per second. This number is almost certainly a lower limit as Gry et al. (1995) using HRS UV spectra of several ions found that $N(H\ I)$ is $\leq 5\times 10^{17}$ cm^2 and could be as low as 3×10^{17} cm^2 along the line of sight to ε CMa. The missing parameters are the electron and protons densities in the line of sight to the star. There are no direct measurements of n_e or n_p for this line of sight, but dispersion measures of radio signal time delays from pulsars provide a good estimate. The mean electron density in the lines of sight to the nearest six pulsars at distances of 156–361 pc is 0.0121 cm^{-3}. With this value of n_e and assuming that $n_p = n_e$, the radius of the Strömgren sphere of ε CMa is $R > 160$ pc. Since the star is only 124 pc away, the LIC and other warm clouds in the CLIC are surrounded by ionized Strömgren sphere gas, which likely has a temperature of 10,000–20,000 K.

This simple calculation has a number of important consequences. First, it explains the hydrogen hole and indicates that the outer edges of the LIC and other clouds facing ε CMa are Strömgren shells as shown in Fig. 4.11. Second, it suggests that the entire Local Cavity may be a Strömgren sphere ionized by ε CMa and other hot stars and white dwarfs. Welsh et al. (2013) computed the sizes of isolated Strömgren spheres surrounding 33 hot white dwarfs inside the Local Cavity based on *EUVE* measurements of their EUV radiation. Since these white dwarfs lie inside of the Strömgren sphere of ε CMa, the combined EUV radiation from all of these stars is sufficient to ionize a region larger than 160 pc and likely the entire Local Cavity.

Is this tentative model of the Local Cavity consistent with other information? One problem is that the thermal pressure in the Local Cavity will be low. If we assume that the total density of the ionized gas is $2n_e$ and the temperature 20,000 K, then $P_{th}/k = 0.024 \times 20,000 = 480$ Kcm^{-3}, which is a factor of 6.6 times smaller than the thermal pressure in the LIC. Since there is no evidence that the LIC is expanding into a vacuum, then other pressure terms should be important. For example, magnetic pressure, radiation pressure, nonthermal gas pressure, cosmic ray pressure, and ram pressure due to turbulent motions could be larger in the Local Cavity than in the LIC.

There is an interesting hybrid scenario that may explain the properties and recent history of the Local Cavity. The isotope ^{60}Fe with a half-life of 2.6 Myr is formed during supernova explosions. This isotope has been discovered in deep ocean ferromanganese crusts that are dated by other isotopes to an age of 1.5−3.2 Myr (Wallner et al., 2016). Supernovae at that time would have created a hot, low density bubble that subsequently cooled by radiation and expansion. The EUV radiation of ε CMa and other hot stars could maintain the hydrogen ionization of the already created but now cooled Local Cavity.

7. Is the interstellar medium static or dynamic?

UV spectra have played an important role in understanding the importance of dynamics in the interstellar medium. Theoretical models of interstellar gas have assumed that the gas is thermally stable, i.e., that the total heating and cooling rates balance locally. These models also assumed that the gas is in pressure equilibrium with its environment so that no hydrodynamic forces can stimulate gas motions. With these assumptions, various authors have shown that interstellar gas naturally separates into regions with distinct temperatures and densities depending on the local heating and cooling rates. For example, the two-component model proposed by Field et al. (1969) consists of cold ($T < 300$ K) and warm ($T \sim 10^4$ K) components. To explain observations of diffuse X-ray emission, later models by McKee and Ostriker (1977) and Wolfire et al. (1995, 2003) also included a hot ($T \sim 10^6$ K) component that fills most of the interstellar volume.

These theoretical models, which do not include hydrodynamic and magnetic forces, predict a nonlinear relation between thermal gas pressure and density. In such models, warm gas clouds like the LIC with densities below 1 cm^{-3} and cold gas clouds like the Leo Cold Cloud with densities above 10 cm^{-3} can coexist in gas pressure equilibrium, but there is an unstable regime at intermediate densities. Jenkins and Tripp (2001) found that typical warm gas cloud thermal pressures are about 2240 Kcm^{-3}, which is in the stable regime. They also found that about 15% of interstellar gas has a thermal pressure greater than 5000 Kcm^{-3} and a very small amount of gas has a thermal pressure above 100,000 Kcm^{-3} with very small size scales (0.01 pc). These results were based on their study of C I fine-structure excitation using the very narrow slit mode of STIS to obtain the highest spectral resolution.

A major problem with the theoretical models is that the thermal gas pressure is generally only a small component of the total pressure. The total pressure in the mid-plane of the Galaxy due to the weight of the overlying material is $P_{total} \approx 22,000$ Kcm^{-3} (Cox, 2005), 10 times the mean thermal pressure. For pressure equilibrium to occur, the total pressure must be balanced by the sum of the magnetic, cosmic ray, ram, and thermal gas pressure terms. The first three pressures are likely larger than the thermal gas pressure. Although difficult to measure, the three nonthermal pressure terms must be included in more realistic future models to explain the range of observed thermal pressures.

deAvillez and Breitschwerdt (2005) developed a three-dimensional magnetohy-drodynamic (MHD) computer code that includes the previously ignored hydrody-namic and magnetic forces and the energy and momentum produced by supernovae. They ran this code with a range of assumed supernova event rates and computed the evolution of the local plasma properties and flows in a grid extend-ing for 10 kpc in the Galactic plane and into the halo. Their simulations include the exchange of matter and energy between the disk and halo with the expansion of hot gas into the halo acting as a pressure relief valve. After about 400 Myr of evolution to remove the input data, their models exhibit nearly a factor of 10^6 range in density and a factor of 10^4 range in pressure. The initially assumed homogeneous magnetic field strength of 3 μG has acquired a range of 0.3–15 μG. If these calculations are realistic, the interstellar medium is very far from pressure and thermal equilibrium.

Fig. 4.12 shows the average values the ram pressure, thermal pressure, and mag-netic pressure as a function of temperature in their simulations. Magnetic pressure

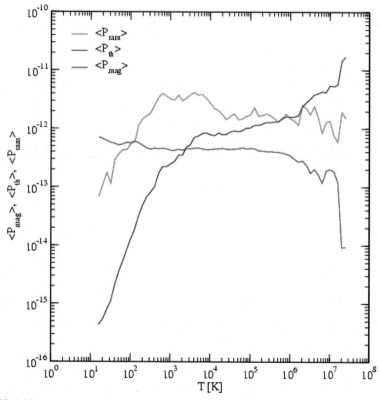

FIGURE 4.12

Comparison of the average ram pressure, P_{ram} (green), thermal pressure, P_{th} (black), and magnetic pressure, P_{mag} (red) as a function of temperature in the Galactic midplane.

Figure from de Avillez, M. A., Breitschwerdt, D., 2005. A&A 436, 585.

dominates over the other pressure terms only for cold ($T < 100$ K) interstellar gas, while thermal pressure dominates only for the hot ($T > 10^6$ K) gas. In the intermediate temperature regime (10^2 to 10^6 K), ram pressure due to turbulent flows dominates the other pressure terms. In these simulations the ISM is highly dynamic with flows that are typically supersonic and superalfvenic. These flows must be included in realistic models of the ISM. Unlike the earlier equilibrium models, there are no distinct phases of the ISM as gas exists at all temperatures, including temperatures that are thermally unstable if equilibrium were assumed. According to the deAvillez and Breitschwerdt (2005) hydrodynamic and MHD models, this intermediate temperature gas occupies most of the volume and represents about half of the mass in the ISM. The hot gas occupies only about 20% of the volume and a negligible fraction of the ISM mass.

What does the local ISM, which consists of warm clouds in the CLIC surrounded by 10,000–20,000 K photoionized gas in the Local Cavity, tell us about the relevance of the equilibrium models and hydrodynamic simulations? The three component theoretical models assume pressure equilibrium between the phases. If the gas temperature in the Local Cavity is indeed 10^6K or close to this temperature, then the thermal pressure, $P_{th}/k = 2n_eT = 2(0.021)10^6 = 42,000$ Kcm^{-3}, can support the weight of overlying mass. However, this thermal pressure would be 10 times that of the LIC, $P_{th}/k = 3,180^{+1,850}_{-1,130}$ Kcm^{-3}. Adding a LIC magnetic pressure with $B \approx 10$ μG would bring these two pressures into balance, but there is no evidence that the magnetic field in the LIC is anywhere this large (Frisch et al., 2011). These pressure discrepancies and the problems with much 10^6 K gas in the Local Cavity argue that the basic assumption of pressure equilibrium among the thermal components in the theoretical models is probably not realistic.

What about the predictions of the dynamic simulations? A potential problem is that in the temperature range 10^2 K to 10^6 K, the simulations predict that P_{ram} should dominate the total pressure. One component of ram pressure is the small spatial scale nonthermal motions in clouds that broaden UV interstellar lines. Typical nonthermal velocities of 2 km/s are subsonic by a factor of four corresponding to a small pressure of about 200 K cm^{-3}. On the other hand, the warm clouds have random cloud velocity differences of 10–15 km/s (Redfield and Linsky, 2008), which are supersonic. The ram pressure associated with the random cloud velocities will be many times P_{th} in the LIC, consistent with the prediction of the simulations. If the Local Cavity gas has a moderate temperature (10,000–20,000 K), as proposed above, then additional pressure terms may be needed in the Local Cavity to balance both the CLIC and the weight of the overlying Galactic mass. In the dynamic simulations, however, pressure imbalances naturally occur and are the sources of flows. Future work based on UV spectroscopy is needed to understand pressure imbalances and their effects on interstellar gas dynamics.

8. Speculations concerning how the interstellar medium may affect the habitability of exoplanets

During its 4.5 Gyr lifetime, the Sun has journeyed through a variety of interstellar environments that could have influenced the evolution of life on Earth, and the changing interstellar environments experienced by exoplanets may stimulate the formation of prebiological molecules and the evolution of putative life forms. It is, therefore, instructive to trace the Sun's interstellar environment back in time to look for different interstellar environments and their possible effects on habitability.

The 26 km/s velocity of the Sun relative to the LIC corresponds to 5.2 AU per year or 25 pc per million years. The Sun likely has been moving through the LIC for about 70,000 years. Since the known warm clouds in the LISM extend to about 15 pc, the Sun has likely been enveloped in a warm partially ionized cloud for most of the past 600,000 years and will likely leave this benign environment in about the same time. During this time inside warm LISM clouds, the heliosphere has shielded the solar system from most high energy cosmic rays. Before entering the LISM, the Sun and its planets traveled through the ionized Local Cavity created by supernova explosions as recent as 1.5–3.2 Myr ago. After leaving the LISM, the Sun will reenter the Local Cavity. Given the present 100–200 pc size of the Local Cavity, both journeys take 4–8 Myr, sufficient time to be influenced by the recent and presumably past supernovae explosions. Before then, the Sun and its cohort of planets would have traversed a number of supernova created bubbles, warm partially ionized clouds and dense molecular clouds including its nascent environment. Could each of these environments have influenced the evolution of life forms?

Müller et al. (2006) computed 27 multifluid heliosphere models treating plasmas and neutral hydrogen self-consistently to address how the heliosphere would react to a diverse range of interstellar environments. The models were computed balancing the total pressure of the solar wind against that of the interstellar medium. The physical processes included in these models are the same as for astrospheres, so the heliosphere models are useful for estimating how interstellar environments could affect exoplanets. When inside of a supernova created bubble of hot fully ionized low density plasma, the heliosphere expands to roughly 3 times its present size. Conversely, when the Sun passes through a dense molecular cloud, simulated by a model with total density 11 cm^{-3} and temperature 100 K, the heliosphere shrinks by a factor of 5 with the heliopause distance only 26 AU rather than 120 AU. The effect of passing through a high velocity cloud would also shrinks the heliosphere.

An important result of shrinking or expanding the heliosphere and astrospheres is to change the modulation of galactic cosmic rays (GCRs). As described by Florinski and Zank (2006) and Müller et al. (2006), the transmission of GCRs through the heliosphere to Earth depends on the magnetic field and turbulence in the heliosphere. In their simulations, an expanded heliosphere will increase shielding thereby decreasing the GCR flux at Earth, and a shrunken heliosphere will likely reduce shielding resulting in greater GCR flux at Earth. Dense molecular clouds where stars form have typical densities of 10^4 cm^{-3}, a factor of 100 times larger than the densest simulation that would shrink the astrosphere much further and further decrease the GCR flux. While these calculations are approximate, they point to important trends of the GCR flux that exoplanets encounter as their host stars traverse the inhomogeneous ISM.

Since high energy radiation, in particular GCRs, can produce mutations in biological matter, an effect of changing interstellar environments is to increase or decrease the rate of biological evolution. In some circumstances enhanced high energy radiation can stimulate the formation of molecules such as amino acids and lipid precursors to complex biological molecules. There are studies that show that stellar NUV radiation between 200 and 280 nm can start the formation of biological precursors through photodissociation of such simple molecules as H^2S to form HS^- and SO_2 to form SO_3^- (Rimmer et al., 2018). Other possible paths for the formation of biological precursors by photodissociation or photoionization with harder radiation including EUV radiation are possible but not yet explored. If such chemical paths can efficiently lead to complex organic molecules, then the ISM environment could be very important. At times when a host star and its exoplanets are enveloped in the fully ionized environment of hot bubbles, EUV radiation from hot stars such as ε CMa and hot white dwarfs can provide a strong EUV flux minimally attenuated by neutral hydrogen in the line of sight through the astrosphere.

Recent studies of radiation effects on biological molecules on the surfaces of exoplanets have explored the effects of NUV (200−280 nm) radiation emitted by their host star. For example, Günther et al. (2020) computed the effects of M dwarf superflares, and concluded that under some circumstances this radiation may be beneficial for the production of ribonucleotides required for RNA synthesis. These studies should be extended to shorter wavelengths to include Ly-α, the brightest emission line in the UV spectrum of M dwarfs (France et al., 2013).

Segura et al. (2010), Youngblood et al. (2017), and Günther et al. (2020) showed that the high energy protons accompanying coronal mass ejections (CMEs), produce nitrogen oxides (NO and NO_2) in exoplanet atmospheres that destroy oxone, the main shield that prevents UV radiation from reaching the surfaces of planets. In their simulations, multiple superflares, which can occur as often as once a day for active M dwarfs, can lead to the permanent depletion of ozone and the sterilization of exoplanet surfaces.

Since GCRs also consist of high energy protons and other nuclei, they may play a role similar to the high energy protons from stellar CMEs. When a host star is in an interstellar environment with minimal neutral hydrogen, such as the Local Cavity that the Sun has and will traverse, the flux of high energy cosmic rays reaching an exoplanet's atmosphere is enhanced significantly. Although the flux of cosmic ray protons is far less than during CME events for a close in exoplanet, the GCR flux is continuous rather than intermittent. Also, the extreme-UV radiation from stars like ε CMa and hot white dwarfs unattenuated by ionized gas in the ISM will penetrate into the atmospheres of exoplanets leading to hydrodynamic escape. Realistic calculations are needed to determine the importance of cosmic rays for exoplanet habitability.

9. Summary

This review covers the roles that ultraviolet observations play in understanding the interstellar medium from the interactions with stellar winds in the heliosphere and astrospheres to the Local Cavity. The review describes the main UV instruments presently in space and proposed for the next decade. The STIS instrument on *HST* plays a critical role in these studies, because high spectral resolution is essential for measuring the composition, physical properties, and kinematic structure of the ISM. A long hiatus between the end of STIS operations and a future high-resolution UV spectrograph would frustrate progress in this field. Atomic hydrogen is both a very useful tool and a hindrance: useful for measuring opacities and mass-loss rates with the astrosphere technique, but an opacity source that obscures the EUV spectrum and important spectral lines of hydrogen, oxygen, and Mg II.

As stars plow through the ISM, interactions of stellar wind protons with neutral hydrogen in the partially ionized interstellar gas produce a structured astrosphere with an astropause separating the stellar wind ions and magnetic field from the interstellar gas and magnetic field. Studies of the heliosphere from satellites in Earth orbit (e.g., *IBEX*) and the *Voyager* spacecraft in the outer heliosphere are measuring the complex phenomena occurring in the Very Local ISM (VLISM) where the inflowing interstellar gas is modified by charge exchange reactions with the thermal and nonthermal components of the solar wind. Stars cooler than spectral type F likely have ionized stellar winds and, therefore, astrospheres that will be similar to the Sun when the surrounding interstellar gas contains neutral hydrogen.

High-resolution spectra allow us to measure the physical and kinematic properties of interstellar gas in the lines of sight to stars. The Sun is located near the edge of the warm partially ionized LIC that is a part of the cluster of local interstellar clouds (CLIC) that extends for perhaps 10−15 pc in all directions. Fifteen of these warm clouds have been identified so far based on their space motions. Whether these clouds are rigid structures confined by magnetic fields, external ionizing radiation, or something else is uncertain at this time.

Surrounding the CLIC and extending for 100−200 pc in all directions is a low density ionized region called the Local Cavity. Initially formed as a hot bubble by supernova explosions, there is a debate as to whether this plasma remains at temperatures of 10^6 K for several million years after the last supernova explosion or whether it has cooled or is now a recombining nonequilibrium plasma. An important clue to answering this question is the presence of very low hydrogen column density in the direction of ε CMa, the star with the highest EUV emission detected by the EUVE satellite. The ionizing flux from ε CMa and hot white dwarfs is sufficient to keep the Local Cavity ionized at a moderate temperature ($T = 10,000 - 20,000$ K). The low electron density needed to explain a Strömgren sphere this large is consistent with dispersion measures of nearby pulsar signals. An inventory of the component pressures in the Local Cavity is needed to test the Strömgren sphere hypothesis.

While stellar UV and X-ray emission especially during superflares and energetic proton emission in CMEs play important roles in determining the habitability of exoplanets, the interstellar environment could also affect the habitability and evolution of life forms on exoplanets. Stars journey through a wide variety of interstellar environments in which variations in the interstellar density can expand or contract their astrospheres. The effect is to change the incident GCR flux, which can lead to the destruction of atmospheric ozone in the same way as the proton flux from CMEs or stimulate the formation of molecules such amino acids and lipid precursors to complex biological molecules. Also, when a host star is enveloped in a region of minimal neutral hydrogen column density such as the Local Cavity, then EUV radiation from hot stars and white dwarfs may erode exoplanet atmospheres continuously rather than by occasional flare events. These speculative scenarios need to be studied to determine for which exoplanets they could be important.

Acknowledgments

The ultraviolet spectra upon which this paper is based were obtained with *HST* Guest Observer programs 11568 and 13332 in collaboration with Prof. Seth Redfield at Wesleyan University. Data were also obtained from the MAST archive of the Space Telescope Science Institute, which is operated by the Association of Universities for Research in Astronomy, Inc. for NASA under contract NAS 5−26555. I thank Seth Redfield for many collaborations over the years and for several figures in this paper.
Facilities: HST(STIS), HST(COS), XMM-Newton, Chandra X-ray Observatory, ROSAT.

References

Boggess, A., Carr, F.A., Evans, D.C., et al., 1978. Nature 275, 372.

Bowyer, S., Malina, R.F., 1991. In: Malina, R.F., Bowyer, S. (Eds.), Extreme Utraviolet Astronomy. Pergamon, New York, p. 333.

Brandt, J.C., Heap, S.R., Beaver, E.A., et al., 1994. PASP 106, 890.

Breitschwerdt, D., Schmutzler, T., 1999. A&A 347, 650.

Bowen, D.V., Jenkins, E.B., Tripp, T.M., et al., 2008. ApJS 176, 59.

Burlaga, L.F., Ness, N.F., Berdichevsky, D.B., et al., 2019. Nat. Astron. 3, 1007.

Capitanio, L., Lallement, R., Vergely, J.L., Elyajouri, M., Monreal-Ibero, A., 2017. A&A 606, A65.

Cox, D.P., 2005. ARA&A 43, 337.

Craig, N., Abbott, M., Finley, D., Jessop, H., Howell, S.B., Mathioudakis, M., Sommers, J., Vallerga, J.V., Malina, R.F., 1997. ApJS 113, 131.

Crutcher, R.M., 1982. ApJ 254, 82.

Davidsen, A.F., Long, K.S., Durrance, S.T., et al., 1992. ApJ 392, 264.

de Avillez, M.A., Breitschwerdt, D., 2005. A&A 436, 585.

Dong, C., Lee, Y., Ma, Y., et al., 2018. ApJ 859, L14.

Edelman, E., Redfield, S., Linsky, J.L., Wood, B.E., Müller, H., 2019. ApJ 880, 117.

Edelstein, J., Bowyer, S., Korpela, E.J., et al., 2001. Ap&SS 276, 177.

Edelstein, J., Korpela, E.J., Adolfo, J., et al., 2006. ApJL 644, L159.

Fitzpatrick, E.I., Massa, D., 1986. ApJ 307, 286.

Fitzpatrick, E.I., Massa, D., 1988. ApJ 328, 734.

Fitzpatrick, E.I., Massa, D., Gordon, K.D., Bohlin, R., Clayton, G.C., 2019. ApJ 886, 108.

Field, G.B., Goldsmith, D.W., Habing, H., 1969. ApJL 155, L149.

Florinski, V., Zank, G.P., 2006. In: Frisch, P.C. (Ed.), Solar Journey. Springer, New York.

France, K., Fleming, B.T., Drake, J.J., et al., 2019. SPIE 11118, 1111808.

France, K., Froning, C.S., Linsky, J.L., et al., 2013. ApJ 763, 149.

Frisch, P.C., Redfield, S., Slavin, J.D., 2011. Ap&SS 49, 237.

Garraffo, C., Drake, J.J., Cohen, O., 2016. ApJ 833, L4.

Green, J.C., Froning, C.S., Osterman, S., et al., 2012. ApJ 744, 60.

Gry, C., Jenkins, E.B., 2014. A&A 567, 58.

Gry, C., Jenkins, E.B., 2017. A&A 598, A31.

Gry, C., Lemonon, L., Vidal-Madjar, A., Lemoine, M., Ferlet, R., 1995. A&A 302, 497.

Günther, M.N., Zhan, Z., Seager, S., et al., 2020. AJ 159, 60.

Hurwitz, M., Bowyer, S., Bristol, R., Van Dyke Dixon, W., Dupuis, J., Edelstein, J., Jelinsky, P., Sasseen, T.P., Siegmund, O., 1998. ApJL 500, L1.

Hurwitz, M., Sasseen, T.P., Sirk, M.M., 2005. ApJ 623, 911.

Jakosky, B.M., Brain, D., Chaffin, M., et al., 2018. Icarus 315, 146.

Jenkins, E.B., Reale, M.A., Zucchino, P.M., Sofia, U., 1996. Ap&SS 239, 315.

Jenkins, E.B., Tripp, T.M., 2001. ApJS 137, 297.

Kondo, Y., de Jager, C., Hoekstra, R., van der Hucht, K.A., Kamperman, T.M., Lamers, H.J.G.L.M., Modisette, J.L., Morgan, T.H., 1979. ApJ 230, 526.

Koutroumpa, D., 2012. Astron. Nachr. 333, 341.

Lallement, R., Bertin, P., 1992. A&A 266, 479.

Lallement, R., Quémerais, E., Bertaux, J.L., Ferron, S., Koutroumpa, D., Pellinen, R., 2005. Science 307, 1447.

Linsky, J.L., 2019. Host Stars and Their Effects on Exoplanet Atmospheres, Lecture Notes in Physics. Springer.

Linsky, J.L., Draine, B.T., Moos, H.W., et al., 2006. ApJ 647, 1106.

Linsky, J.L., Redfield, S., Tilipman, D., 2019. ApJ 886, 41.

Lyu, C.H., Bruhweiler, F.C., 1996. ApJ 459, 216.

Malamut, C., Redfield, S., Wood, B.E., Ayres, T.R., 2014. ApJ 787, 75.

Mason, K.O., Breeveld, A., Much, R., Carter, M., Cordova, F.A., Cropper, M.S., Fordham, J., Huckle, H., Ho, C., 2001. A&A 365, L36.

Mason, K.O., Breeveld, A., Hunsberger, S.D., James, C., Kennedy, T.E., Roming, P.W.A., Stock, J., 2004. SPIE 5165, 277.

McKee, C.F., Ostriker, J.P., 1977. ApJ 218, 148.

Möbius, E., Bzowski, M., Frisch, P.C., et al., 2015. ApJS 220, 24.

Moos, H.W., Sonnebron, G., 2006. Astrophysics in the far utraviolet: five years of discovery with FUSE. In: Sonneborn, G., Moos, H.W., Andersson, B.-G. (Eds.), ASP Conf. Series, vol. 348, p. 4.

Morrissey, P., Schiminovich, D., Barlow, T.A., et al., 2005. ApJL 619, L7.

Morton, D.C., 1967. ApJ 147, 1017.

Müller, H.-R., Florinski, V., Heerikhuisen, J., et al., 2008. A&A 491, 43.

Müller, H.-R., Frisch, P.C., Florinski, V., Zank, G.P., 2006. ApJ 647, 1491.

Peek, J.E.G., Heiles, C., Peek, K.M.G., Meyer, D., Lauroesch, J.Y., 2011. ApJ 735, 129.

Redfield, S., Falcon, R.E., 2008. ApJ 683, 207.

Redfield, S., Linsky, J.L., 2008. ApJ 673, 283.

Redfield, S., Linsky, J.L., 2015. ApJ 812, 125.

Rimmer, P.B., Xu, J., Thompson, S.J., 2018. Sci. Adv. 4, eaar3302.

Rogerson, J.B., Spitzer, L., Drake, J.F., Dressler, K., Jenkins, E.B., Morton, D.C., York, D.G., 1973. ApJL 181, 97.

Schlafly, E.F., Meisner, A.M., Kainulainen, J., et al., 2016. ApJ 821, 78.

Segura, A., Walkowitz, L.W., Meadows, V., Kasting, J., Hawley, S., 2010. Astrobiology 10, 751.

Shustov, B., Gómez de Castro, A.I., Sachkov, M., et al., 2018. Astrophys. Space Sci. 363, 62.

Slavin, J.D., Frisch, P.C., 2008. A&A 491, 53.

Strömgren, B., 1939. ApJ 89, 526.

Strub, P., Krüger, H., Sterken, V., 2015. ApJ 812, 140.

Vallerga, J., 1996. SSRv 78, 277.

Vallerga, J., 1998. ApJ 497, 921.

Vidotto, A.A., Jardine, M., Opher, M., Donati, J.F., Gombosi, T.I., 2011. MNRAS 412, 351.

Wallner, A., Feige, J., Kinoshita, N., et al., 2016. Nature 532, 69.

Welsh, B.Y., Shelton, R.L., 2009. Astrophys. Space Sci. 323, 1.

Welsh, B.Y., Wheatley, J., Dickinson, N.J., Barstow, M.A., 2013. PASP 125, 644.

Wolfire, M., McKee, C., Hollenbach, D., Tielens, A., 1995. ApJ 453, 673.

Wolfire, M., McKee, C., Hollenbach, D., Tielens, A., 2003. ApJ 587, 278.

Wood, B.E., Müller, H.R., Witte, M., 2015. ApJ 801, 62.

Wood, B.E., Redfield, S., Linsky, J.L., Müller, H.-R., Zank, G.P., 2005a. ApJS 159, 118.

Wood, B.E., Müller, H.-R., Zank, G.P., Linsky, J.L., Redfield, S., 2005b. ApJL 628, L143.

Wood, B.E., Müller, H.-R., Redfield, S., Edelman, E., 2014. ApJL 781, L33.

Woodgate, B.E., Kimble, R.A., Bowers, C.W., et al., 1998. PASP 110, 752.

Youngblood, A., France, K., Parke Loyd, R.O., et al., 2016. ApJ 824, 101.

Youngblood, A., France, K., Parke Loyd, R.O., et al., 2017. ApJ 843, 31.

Zank, G.P., Heerikhuisen, J., Wood, B.E., Pogorelov, N.V., Zirnstein, E., McComas, D.J., 2013. ApJ 763, 20.

UV facilities for the investigation of the origin of life

Ana I. Gómez de Castro[1,2], Martin A. Barstow[3], Noah Brosch[4], Patrick Coté[5], Kevin France[6], Sara Heap[7], John Hutchings[5], S. Koriski[4], Jayant Murthy[8], Coralie Neiner[9], Aki Roberge[10], Julia Román-Duval[11], Jason Rowe[12], Mikhail Sachkov[1,13], Evgenya Schkolnik[14], Boris Shustov[1,13]

[1]*Joint Center for Ultraviolet Astronomy (JCUVA), Universidad Complutense de Madrid, Madrid, Spain;* [2]*U.D. Astronomia y Geodesia, Facultad de CC Matemáticas, Universidad Complutense de Madrid, Madrid, Spain;* [3]*School of Physics and Astronomy, University of Leicester, Leicester, United Kingdom;* [4]*Tel Aviv University, Tel Aviv, Israel;* [5]*National Research Council, Herzberg Astronomy and Astrophysics Research Centre, Victoria, BC, Canada;* [6]*Laboratory for Atmospheric and Space Physics, University of Colorado, Boulder, CO, United States;* [7]*University of Maryland, College Park, MD, United States;* [8]*Indian Institute of Astrophysics, Bangalore, Karnataka, India;* [9]*Observatoire Paris-Meudom, Meudom, France;* [10]*Goddard Space Flight Center — NASA, Greenbelt, MD, United States;* [11]*Space Telescope Science Institute, Baltimore, MD, United States;* [12]*Bishop's University, Sherbrooke, QC, Canada;* [13]*Institute of Astronomy of the Russian Academy of Sciences, Moscow, Russia;* [14]*School of Earth and Space Exploration, Arizona State University, Phoenix, AZ, United States*

1. Introduction

Ultraviolet (UV) observatories are powerful tools to study the Universe and, particularly those processes associated with the origin of life and the chemical enrichment that makes it possible. Though the first UV observations of the sky date back to 1957 (Byram et al., 1957), a true worldwide community of UV specialists was only made feasible by the *International Ultraviolet Explorer* (IUE), a small observatory by today's standards that was operational during 17 years granting access to UV data to generations of astronomers (see Brosch, 1998 for a review on the early days of UV astronomy). The golden decade for UV observatories started just after IUE, in 2000–10, when three missions, two of which were fully dedicated to UV astronomy, were flying at the same time: the *Far-Ultraviolet Spectroscopic Explorer* (FUSE), the *Galaxy Evolution Explorer* (GALEX) and the *Hubble Space Telescope* (HST). Currently, only Hubble, a 30-year-old mission, keeps flying and acquiring fundamental data for all areas of astrophysics. High-energy astrophysics observatories such as XMM-Newton and SWIFT also incorporate small UV cameras but their use for astrobiological purposes is rather limited.

Ultraviolet Astronomy and the Quest for the Origin of Life. https://doi.org/10.1016/B978-0-12-819170-5.00004-X

HST aging and the scientific demands for access to the UV range are pushing forward a next generation of instrumentation. There are missions already flying like ASTROSAT-UVIT, missions that will fly in a few years (Spektr-UF/WSO-UV and the Chinese Space Station Telescope) but the bulk is coming to be flown in the 2030—2040 decade with some cubesat and small-sat precursors acting as technological testers while granting access to the UV sky for specific scientific purposes.

This contribution collects inputs from most of these missions and projects concerning their characteristics and their foreseen impact in astrobiological research. The inputs are ordered in three sections: flying missions, under-construction missions, and projects/proposals. Each project has been awarded two to four pages. At the end, a comprehensive summary is provided including the relevant information for future reference.

2. Operational missions

2.1 Hubble space telescope

Summary provided by Julia Román-Duval on behalf of the Hubble space telescope

The HST, named after astronomer Edwin Hubble, is an international collaboration between the National Aeronautics and Space Administration (NASA) and the European Space Agency (ESA). Launched to low-Earth orbit in 1990 with space shuttle Discovery, this versatile observatory, equipped with a 2.4 m mirror, imagers, and spectrographs operating from the UV to the near-infrared, has transformed the field of astronomy, enabling many scientific breakthroughs across astrophysics. In tandem with efforts at other observatories, Hubble is famous for helping demonstrate the accelerating expansion of the universe and the current tension with other measurements that may be pointing to new physics. A few other highlights include observations of the growth of galaxies through cosmic times; observations of black holes revealing their prevalence throughout the universe; transit spectroscopy of exoplanets revealing their atmospheric compositions; and observations of solar system objects, for example, revealing subsurface oceans on Jupiter's moon Ganymede. To date, discoveries based on 1.4 million Hubble observations have led to more than 17,000 peer-reviewed publications. Hubble observing is open to the astronomical community around the world via the Guest Observer program. Opportunities to request observing time come on a variety of scales, including a large annual call, and multiple opportunities on a more frequent cadence to address new discoveries and transient phenomena.

The success of Hubble in part comes from its serviceable design: five servicing missions executed with the space shuttles between 1993 and 2009 have repaired, upgraded, or replaced critical hardware and the instruments. The five instruments onboard HST for the remainder of the mission are the Cosmic Origins Spectrograph (COS, installed during servicing mission 4—SM4—in 2009), the Wide-Field Camera 3 (WFC3, also installed in 2009), the Advanced Camera for Surveys (ACS, installed during SM3b in 2002 and repaired during SM4 in 2009), the Space

Telescope Imaging Spectrograph (STIS, installed during SM2 in 1997 and also repaired during SM4), and the Fine Guidance Sensor (FGS, the only original instrument onboard Hubble since 1990). We will focus here on UV-capable instruments: COS, STIS, WFC3, and ACS.

2.1.1 The Cosmic Origins Spectrograph (COS)

The Cosmic Origins Spectrograph (COS) is the most sensitive UV spectrograph ever built and flown. With a 2.5″ aperture, it is optimized for point sources.

The COS FUV channel includes two side-by-side windowless $16,384 \times 1024$ pixel cross-delay line microchannel plate (MCP) detectors with very low background ($\sim 3 \times 10^{-6}$ cts/pix/s dark current rate). The FUV detectors are illuminated by photons from the 2.5″ diameter primary science aperture (or, very rarely, the bright object aperture) passing through three possible gratings: two medium-resolution gratings ($R \sim 20,000$), the G130M (900–1470 Å) and G160M (1340–1800 Å); and a low-resolution grating ($R \sim 3000$), G140L (770–2150 Å). The wavelength range of the G130M grating is covered through eight central wavelength settings (or "cenwave"), while six cenwaves are included in the G160M and three cenwaves in the G140L. An exposure can be taken one cenwave at a time. The full FUV wavelength span can be covered with three cenwave settings. A grating can be selected by rotating the Optical Select Mechanism (OSM).

The COS NUV channel is comprised of one 1024×1024 Cesium Telluride multianode microchannel array (MAMA), which is the spare NUV detector from the STIS (see later in the chapter). The background rate for the NUV channel is substantially higher than that for the FUV, with a dark rate of $\sim 9 \times 10^{-4}$ cts/pix/s. Incoming photons are directed to the NUV detector via the OSM. Photons pass through four possible gratings: three medium-resolution gratings G185M (1670–2125 Å), G225M (2070–2525 Å), and G285M (2480–3225 Å); and one low-resolution grating, G230L (1340–3360 Å). Due to rapidly declining throughput, the G285M is practically not used any longer. The wavelength coverage for each grating is determined by different cenwave settings. The optical path of photons through the NUV channel results in a spectrum being split into three distinct stripes on the NUV detector. For the medium-resolution gratings, each stripe covers about 30 Å, while the wavelength gaps between stripes are also about 30 Å wide. As a result, the full continuous NUV wavelength coverage requires the use of five cenwaves. The NUV detector also provides imaging capability, which is used for target acquisitions and science.

While Hubble is famous for its beautiful images, its spectroscopic capabilities in the UV have enabled ground-breaking discoveries in the quest for the origins of life. COS has led to recent breakthroughs in our understanding of the circumgalactic medium (CGM) and the lifecycle of gas and metals in galaxies and their halos. Our understanding of the baryon cycle is complemented by UV spectroscopy of massive stars in nearby galaxies (e.g., the Magellanic Clouds, low-metallicity galaxies in the Local Volume) and clusters in low-redshift galaxies, revealing the chemical composition of young stars and their surrounding interstellar medium.

In the Milky Way, COS has enabled measurements of the composition of protoplanetary disks (e.g., HI, H_2, and CO), and planet atmospheres (via transit spectroscopy). Planetary debris rich in water and exhibiting the chemical signatures of the building blocks of Earth-sized rocky planets have been discovered with COS around a few white dwarfs. On yet smaller scales in the Solar System, COS has revealed the composition of asteroids and comets, providing the water production rate for the latter.

2.1.2 The space telescope imaging spectrograph (STIS)

Prior to the installation of COS in 2009, STIS was the spectroscopic workhorse of HST. STIS suffered a power supply failure in 2004 and was repaired during Servicing Mission 4 in 2009.

With NUV and FUV MAMAs and a charge coupled device (CCD), all 1024×1024 in size, STIS integrates all possible capabilities in one instrument: imaging and spectroscopy from the FUV to the NIR, and even a coronographic mode. We will focus here on the UV capabilities and science for STIS.

The FUV and NUV channels can be illuminated by variety of gratings: low ($R \sim 1000$) and medium-resolution ($R \sim 15,000$) first order spectroscopy gratings (G140L and G140M in the FUV covering 1150–1750 Å; G230M and G230L in the NUV covering 1600–3200 Å); medium ($R \sim 40,000$) and high ($R \sim 100,000$) resolution echelle gratings (E140M, E140H in the FUV; E230M, E230H in the NUV). Except for the E140M grating, which continuously covers the 1150–1700 Å range, all medium and high-resolution gratings make use of several central wavelength settings to cover the FUV and NUV, with one cenwave being selected for each exposure. While COS is optimized for point sources, STIS is capable of performing spectral imaging of extended targets thanks to its long slits and prism. Both NUV and FUV imaging are also available.

The UV spectroscopic sensitivity of STIS is substantially lower than COS (see Fig. 5.1) and the dark current rate is higher (1×10^{-3} cts/pix/s for the NUV, 1×10^{-5} cts/pix/s for the FUV). However, bright object limits are less stringent for STIS than for the very sensitive COS spectrograph. This makes the two spectrographs very complementary: bright targets, such as massive stars in the Milky Way and Magellanic Clouds, could be too bright for COS, but can be observed efficiently with STIS. On the faint end, dim targets, such as quasars, stars outside the Milky Way and Magellanic system, low-mass stars in the Milky Way, are better observed with COS.

The range of discoveries enabled by STIS in the UV is as broad as its capabilities. STIS was instrumental to our understanding of the role supermassive black holes play in galaxy formation and evolution by demonstrating that they are ubiquitous in the universe. In the Milky Way and nearby galaxies, STIS has led to measurements of the composition of stars, planetary nebulae, and HII regions. Its spectral imaging capability has uncovered the structure and composition of one of the most massive stellar systems in our Galaxy, Eta Carinae. STIS spectroscopy of M stars, thought to harbor habitable planets, has constrained their UV radiation field, its effect on the atmosphere evaporation of potential planets (Fig. 5.3), and on the

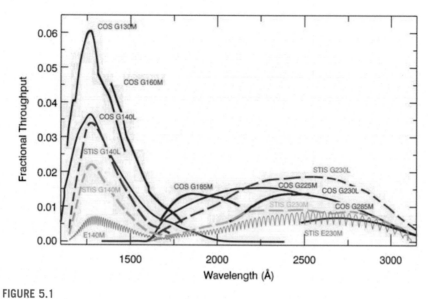

FIGURE 5.1

Fractional throughput of the COS and STIS UV modes as a function of wavelength.

From the HST primer.

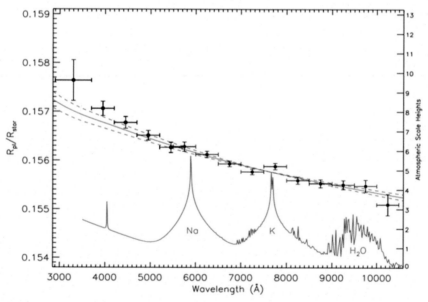

FIGURE 5.2

Transmission spectrum of HD189733b measured by ACS and STIS. Model predictions for Rayleigh scattering by haze is shown as red lines. The haze-free model is shown in pink, and is clearly inconsistent with the observations.

From Sing et al. (2011).

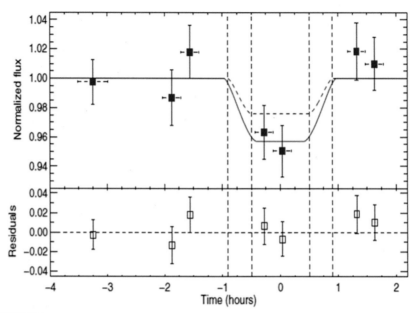

FIGURE 5.3

Flux in the hydrogen Lyman-α line recorded by STIS during the transit of H189733b. The ingress and regress of the transit are indicated by the vertical *dashed lines*. A simple occultation model is shown as a *solid line* (*dashed line* for the optical). The dip in Lyman-α flux during the transit indicates the presence of a hydrogen cloud escaping the atmosphere of the planet, a sign of evaporation due to the UV radiation of the very close host star.

From Bourrier et al. (2013).

conditions for habitability. Time series transit STIS spectroscopy of exoplanets in the UV has revealed key elements of their atmospheric composition, and the presence of dominant high-altitude haze (Fig. 5.2). Closer in the solar system, STIS obtained measurements of the composition of comets (the abundance of the OH molecule, for example), planets, and moons. Water plumes on Europa were discovered in FUV images of Jupiter's moon transiting over the face of its host planet, while the structure of Auroras on Jupiter and Saturn revealed the interaction between the Sun's wind and the gaseous planets' magnetic field. The modeling of the dynamics of auroral ovals observed on Ganymede with STIS in the UV led to the discovery of subsurface oceans on the largest moon of the Solar System.

2.1.3 Wide-field camera 3 (WFC3) and advanced camera for surveys (ACS)
The imagers on HST, WFC3 and ACS also have interesting capabilities in the UV.

The WFC3/UVIS channel is equipped with a diverse set of NUV filters and a NUV grism (G280) for NUV-optical low-resolution spectral imaging (R~70 at 3,000 Å). While the WFC3/UVIS throughput does not extend below 200 nm, it

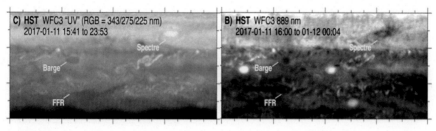

FIGURE 5.4

UV (left) and NIR (right) images of Jupiter taken by WFC3, showing the appearance of cyclones at different wavelengths. These observations can constrain the effects of cyclone dynamics on haze.

From Wong et al. (2020).

has a much wider field-of-view (162″ × 162″) than STIS (25″ × 25″), and a higher throughput, making it the instrument of choice for NUV imaging.

ACS is complementary to WFC3, covering the FUV through its solar-blind channel (SBC, 115–170 nm). The SBC is equipped with several FUV filters and two prisms (covering 1250–2000 Å with R ∼ 100 at 1300 Å). ACS/SBC has a wider field-of-view (34″ × 31″) than STIS, albeit at slightly reduced resolution (pixel scales of 0.032″ for ACS vs. 0.025″ for STIS), and a higher throughput, making it the preferred instrument for FUV imaging.

The UV filters on WFC3 and ACS have been used in many observational studies of star-formation, stellar populations, and interstellar dust in the nearby universe. In the Solar System, UV imaging with WFC3 and ACS enabled studies of aurorae and weather on gaseous planets in the Solar System (Fig. 5.4). Furthermore, the WFC3 grism and ACS prisms have expanded the samples of transit spectra of exoplanets, constraining the composition of their atmosphere and the presence of clouds and hazes (see Fig. 5.2).

2.2 The ultraviolet imaging telescope (UVIT) on board the ASTROSAT mission

Summary provided by Jayant Murthy and John Hutchings

The AstroSat satellite was launched by ISRO into a near-equatorial orbit in September 2015 with a multiwavelength complement of instruments, most in the X-ray. One of the primary instruments is the Ultraviolet Imaging Telescope (UVIT) which consists of two coaligned telescopes, each with a spatial resolution of close to 1″ over a 30′ field-of-view. The telescopes feed three microchannel-based detectors, which are identical except for their photocathodes. One telescope is dedicated to the far-ultraviolet (FUV) with a passband of 125–179 nm while the other feeds a near-ultraviolet detector (194–304 nm) and a visible detector (385–530 nm), with the different bands separated by a dichroic. Each detector includes a filter wheel with narrow-band filters intended to observe in astrophysically important spectral lines. A grating allows low-resolution spectroscopy over the entire field. The UVIT instrument complements the GALEX in that it provides much finer detail over a smaller field-of-view. The FUV detector has about half the sensitivity of the Galaxy Evolution Explorer, because of the smaller primary mirror of the UVIT mirror, while the NUV detector has about the same sensitivity as GALEX, because of the increased efficiency of the UVIT dichroic compared to that of GALEX (see Fig. 5.5). The primary purpose of the visible sensor has been to improve the registration of images. There has been an excellent science return from Astrosat with UVIT observing extended regions from supernova remnants to star-formation regions in other galaxies with most of the results coming from the morphology of the objects in the UV with a higher resolution than previously available.

The primary promise of UVIT in the search for extraterrestrial planets and life is in using its high cadence (30 Hz). This may be useful in conducting a survey of M dwarfs which might provide a good environment for life if it were not for the intense flares. UVIT may also be able to observe eclipsing planets and, indeed, such attempts have been made. As UVIT completes its survey of the sky, serendipitous observations of such stars may provide useful limits to the search for habitable planets.

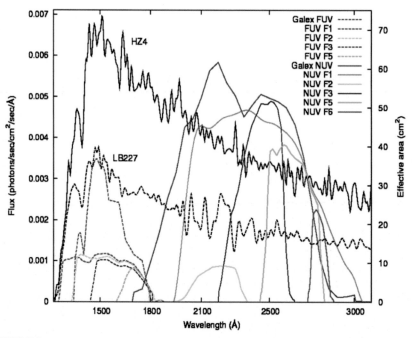

FIGURE 5.5

Plots of standard stars spectra with over plotted effective areas of UVIT and GALEX filters.

3. Missions under construction

3.1 Spektr-UV/world space observatory—ultraviolet (WSO-UV)

Summary provided by Ana I. Gómez de Castro, Boris Shustov, Mikhail Sachkov on behalf of the Spektr-UF (WSO-UV) team

Spektr-UF (or the World Space Observatory—UV) is a middle size space telescope devoted to the ultraviolet observation of astronomical and Solar System sources. The project is part of the Russian scientific space program and will be launched in 2025, 6 years after the successful launch of its predecessor Spektr-RG.

The scientific payload consists of a 170-cm primary space telescope, equipped with instrumentation for imaging and UV spectroscopy in the 115—315 nm range, basically from Lyman-α (Lyα) to the atmospheric cut-off. The WSO-UV will be placed in geosynchronous orbit in 2025 by an Angara A5 launcher from the Vostochny Cosmodrome becoming the first 2-m class UV observatory flown into High Earth Orbit (HEO).

Main instruments are the spectrographs unit (WUVS) and field camera unit (FCU). In addition, the FGS operating at optical (R-Johnson band) wavelengths is susceptible to be used for astronomical purposes. WUVS is being provided by Russia and FCU results from a collaboration between Russia and Spain. Spain is also involved in the scientific operations and ground segment of the project.

WSO-UV instrumentation is designed to provide:

- Spectroscopic observations in the 115—315 nm range with dispersion 50,000 (see Fig. 5.6 for a comparison between the expected sensitivity of WSO-UV spectrographs and those on Hubble).
- Long slit spectroscopy with dispersion 1,000 and angular resolution 0.5 arcsec.
- Imagery of space objects in the 115—176 nm range with angular resolution (up to 0.1 arcseconds) and slitless spectroscopic capabilities, and wide-field imaging in the 175—600 nm range.

3.1.1 WSO-UV capabilities for the investigation of the origin of life

The WSO-UV science program, or core program, considers the investigation about the origin of life in the Universe a priority. Core program observations are to be carried in the first 2 years of the mission and the first call was issued in 2018. The observatory will also host other scientific programs such as the national programs of the countries involved in the project and an open program to the international astronomical community (about 40% of the time after the second year of the mission will be awarded to this program).

High dispersion UV spectrographs are very well suited for the investigation of the chemical abundances in the interstellar, circumstellar and circumgalactic medium.

The low dispersion spectrograph is equipped with a long slit (1 arcsec wide, 72.8 arcsec long) to measure the UV radiation from extended sources such as jets, nebulae, protoplanetary disks and to study the extinction law in photoionized star-

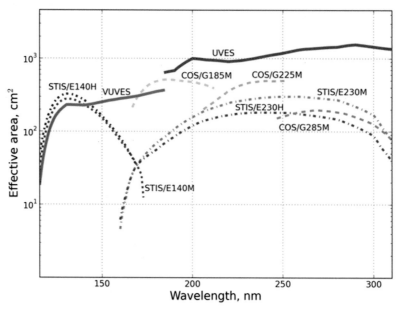

FIGURE 5.6

The sensitivity of WSO-UV spectrographs (WUVS) compared with those in Hubble.

forming regions to actually determine dust grain growth and PAH depletion in these environments (see i.e., Gómez de Castro et al., 2015). The observations of comets and the determination of the photoevaporation rate from the ices and the photodissociation rate of important cometary compounds such as H_2O or CO will be measurable with WSO-UV (Shustov et al., 2018).

The FCU has two channels: near-UV (NUV) and far-UV (FUV).

The NUV channel operates from 175 nm to longer wavelengths. As the detector is a CCD, the system is sensitive up to 800 nm though it is optimized for the 175–315 nm range; the peak efficiency is 48% at 230 nm. The NUV channel has a very wide-field of view (10 × 7.5 arcmin) observed with an angular resolution of 0.15 arcsec hence, it is very well suited to monitor transits and flares in exoplanetary systems over wide fields. A detailed description of the filter set for this channel can be found in Sachkov et al. (2020).

The FUV channel works from 115 to 175 nm and it is equipped with a high sensitivity MCP- type detector. The detector may be operated in photon counting mode to monitor rapid flux variations and flares and has a modular gain designed to increase the dynamical range for certain purposes. The field-of-view is circular with diameter 1.40 arcmin and the angular resolution is 0.1 arcsec. Slitless spectroscopy is feasible with two prisms tuned to Lyα and CIV wavelengths. The prism dispersion decays rapidly with wavelength; for instance, the Lyα prism provides a dispersion of 1000 at Lyα (122 nm) that drops conveniently to 650 at the wavelength of the O I lines at 130 nm, very relevant for astrobiological research. The CIV prism provides a dispersion of 1300 at the SiIV transition (140 nm), 700 at CIV (155 nmn), 650 at CI (161 nm) and 600 at He II (160 nm). FUV spectroscopy of the photoevaporating atmospheres of giant planets, or protoplanetary disks will be feasible provided the safety constrains of the detector are met.

There is a new instrument possibly coming to WSO-UV, the *UV Spectrograph for observation of Earthlike Exoplanets* (UVSPEX) funded by JAXA. UVSPEX is a very efficient UV spectrograph designed to provide UV spectra in the 115−175 nm range with dispersion 240 at Lyα to study exoplanets exospheres.

3.2 CUTE: a cubesat mission for ultraviolet transit spectroscopy of short-period exoplanets

Summary provided by Kevin France on behalf of the CUTE team

Atmospheric escape is a fundamental process that affects the structure, composition, and evolution of many extrasolar planets. It is known to have shaped the early atmospheres of Venus, Earth, and Mars, which subsequently followed different evolutionary paths. While the rapid hydrodynamic escape that is believed to have influenced solar system planets does not operate today, strong signatures of atmospheric loss can be observed on short-period extrasolar planets (e.g., Vidal-Madjar et al., 2003). Owing to their large sizes and short-periods, the physics of atmospheric mass-loss on giant exoplanets can be studied with a modest-sized instrument, operating with long effective observing windows, in the ultraviolet part of the spectrum. The Colorado Ultraviolet Transit Experiment (CUTE) was designed to spectrally isolate diagnostic atomic and molecular transitions arising in these atmospheres to study their composition and the physics of atmospheric mass-loss.

CUTE will conduct atmospheric transmission spectroscopy of short-period, giant planets in the near-ultraviolet (NUV; 255−330 nm), where the stellar flux is two to three orders of magnitude higher than in the FUV band, and transit light-curves can be measured against a more uniform, photospheric stellar surface intensity distribution (Llama and Shkolnik 2015). The CUTE NUV bandpass includes the Fe II complex near 260 nm and Mg II doublet at 279.6/280.3 nm that are observed on several Hot Jupiters (see Sing et al., 2019). This NUV band also contains a continuum that probes scattering by high-altitude clouds and hazes.

The CUTE payload is a magnifying NUV spectrograph fed by a rectangular Cassegrain telescope. The spectrogram is recorded on a back-illuminated, UV-enhanced e2v CCD42-10 that is maintained at a nominal operating temperature by the combination of passive cooling through a radiator panel and a thermal electric cooling system. CUTE employs the 6U Blue Canyon Technology XB1 bus to provide critical subsystems including power, command and data handling, communications, and attitude control. The CUTE aperture is a 206×84 mm, F/0.75 primary mirror that is part of an F/2.6 Cassegrain telescope. The rectangular shape of the primary is enabled by the long axis of the 6U CubeSat chassis and allows for three times more throughput than possible with a 1U circular aperture (Fleming et al., 2018).

The hyperbolic secondary mirror is cantilevered off of the primary mirror by means of an invar tower. A compact (15×6 mm) fold mirror redirects the beam 90° through a slit ($80'' \times 1400''$ projected) at the Cassegrain focus. Once through the slit, the starlight is diffracted, redirected, and magnified by a spherical, aberration correcting, ion-etched holographic grating. A second fold mirror provides additional aberration corrections before the beam reaches the CCD. The optics, detector, custom avionics, and thermal electric cooling system were tested and flight ruggedized by the CUTE team prior to instrument integration (see Fig. 5.7).

All telescope mirrors are coated with Al + MgF$_2$ to prevent the formation of an oxide layer, while the grating has a bare Al coating. Preflight efficiency and environmental testing showed no measurable loss in Al grating efficiency over time. The CUTE science payload was assembled and calibrated at the University of Colorado in 2019 and 2020. Spectrograph focus tests were performed, resulting in a preflight resolving power estimate of $R = \lambda/\Delta\lambda \approx 2{,}000$, which provides ample separation of the relevant atomic, molecular, and continuum bands in this range. The effective area of the CUTE instrument is 20–25 cm^2 across the NUV band, validated by component-level testing, sufficient to execute CUTE's baseline science mission.

CUTE is scheduled to launch as a secondary payload on NASA's LANDSAT-9 mission in September 2021. Communications and data downlink are managed directly by the University of Colorado. The student-led CUTE mission operations team is mentored by professional staff to support communications and science data downlink for the mission. Following launch and the establishment of communication with the spacecraft, the mission operations team conducts instrument health

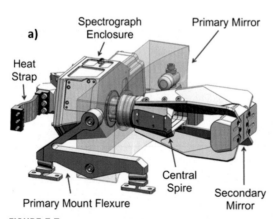
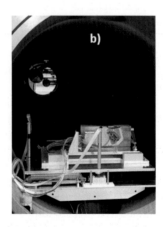

FIGURE 5.7

The left panel shows a schematic with the optomechanical hardware labeled. The CCD focal plane assembly (occulted in this view) is mounted at the back of the spectrograph enclosure. The right panel pictures CUTE during end-to-end testing in the UV calibration facilities at the University of Colorado.

Adapted from (Egan et al., 2018) and (France et al., 2019).

checks, brings up mission critical subsystems (e.g., TEC) over multiple orbits, and performs payload-to-star tracker alignment updates before transitioning into normal science operations mode. Two early release science targets are observed during the transition from commissioning to normal operations, which will mark the beginning of the primary CUTE science mission. CUTE's dedicated mission architecture enables it to survey approximately a dozen short-period planets in an 8 months science mission, acquiring in- and out-of-transit spectroscopy of 6−10 transits for each of the target systems.

3.3 Monitoring the high-energy radiation environment of exoplanets around low-mass stars with SPARCS (star-planet activity research CubeSat)

Summary provided by Evgenya Shkolnik on behalf of the SPARCS team

Tens of billions of M dwarfs (0.1−0.6 $M_\odot$) in our galaxy host at least one small planet in the habitable zone (HZ; e.g., Dressing and Charbonneau 2015). The stellar UV radiation upon the planets from M dwarfs is strong and highly variable, and impacts planetary atmospheric loss, composition and habitability. These effects are amplified by the extreme proximity of their HZs ($\sim$0.1−0.4 AU; Kopparapu 2013).

The James Webb Space Telescope (JWST) and upcoming extremely large ground-based telescopes (ELTs) aim to characterize HZ M dwarf planets and attempt the first spectroscopic search for life beyond the Solar System. Knowing the UV environments of M dwarf planets will be crucial to understanding their atmospheric composition and a key parameter in discriminating between biological and abiotic sources for observed biosignatures (Meadows 2017). The UV flux emitted during the superluminous premain sequence phase of M stars drives water loss and photochemical O_2 buildup for terrestrial planets within the HZ (Luger and Barnes 2015). This phase can persist for up to a billion years for the lowest mass M stars (e.g., Shkolnik and Barman, 2014; Schneider and Shkolnik, 2018). Afterward, UV-driven photochemistry during the main sequence phase strongly affects a planet's atmosphere (e.g., Segura et al., 2010; Tilley et al., 2019), could limit the planet's potential for habitability, and may confuse studies of habitability by creating false chemical biosignatures (Davis et al. in prep).

The SPARCS observatory will be the first mission to provide the time-dependent spectral slope, intensity, and evolution of M dwarf stellar UV radiation over time-scales of days, weeks, and months. These measurements are crucial to interpreting observations of planetary atmospheres around low-mass stars (Fig. 5.8).

SPARCS will be a 6U CubeSat devoted to monitoring $\sim$25 M stars in two UV bands: SPARCS far-UV (S-FUV: 153−171 nm) and SPARCS near-UV (S-NUV: 260−300 nm). It will house a 9 cm telescope with a 0.7′ field-of-view (FOV). For each target, SPARCS will observe continuously between one and three complete stellar rotations (4−45 days) over a mission lifetime of at least one year. SPARCS bands (see Fig. 5.9) include emission lines formed at various stellar heights to measure variability, guide the upper-atmospheric stellar models and predict the

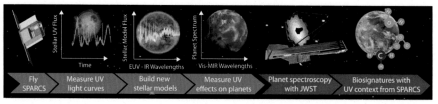

FIGURE 5.8

The flowchart above demonstrates the science progression from SPARCS data to stellar and planetary modeling to informing JWST and other future ground and space telescope observations and data interpretation.

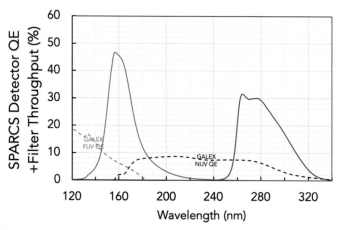

FIGURE 5.9

SPARCS detector quantum efficiency + throughput of the two SPARCS filters compared to the GALEX NUV and FUV detector quantum efficiency (Nikzad et al., 2012, Jewell et al., 2018).

extreme-UV flux (Peacock et al., 2019, 2020). These empirically guided stellar models will provide the exoplanetary community with the much-needed input spectra for time-dependent photochemical, climate and atmospheric escape models for terrestrial exoplanets. In turn, these planet models will inform observations to be performed with JWST, ELTs, and future missions, and allow for accurate interpretation of exoplanetary spectra.

SPARCS enables a wide array of ancillary investigations as its large 40″ FOV will have other UV varying objects. One example is the UV monitoring of active galactic nuclei (AGN), a probe of accretion of mass by a supermassive black hole at the center of its galaxy (Peterson 1993, Gorjian et al. in prep.). AGNs have been studied by all UV-capable satellites, however, only a few of those observations were dedicated to variability and almost none to short cadence monitoring. SPARCS will perform the monitoring necessary to increase our understanding of accretion disks around supermassive black holes. The SPARCS target FOVs include eight AGN bright enough for this experiment.

SPARCS will advance UV detector technology by flying high quantum efficiency (QE), UV-optimized detectors developed at JPL (Nikzad et al., 2012), providing >5× the sensitivity of the detectors used by GALEX, the most recent dedicated UV mission. SPARCS will pave the way for their possible application in a future flagship and Explorer-class missions.

4. Projects

The following is a list of active projects that includes proposals over a wide variety of physical and budgetary scales from the Large Ultraviolet Optical Infrared Surveyor (LUVOIR) propose to become the next NASA flagship, to moderate probe-scale missions down to small cubesats. The list is not complete, especially at the low-mass end, and the projects are sorted by their size.

4.1 The large UV/Optical infrared surveyor (LUVOIR) mission concept

Summary provided by Aki Roberge and the LUVOIR study team

The LUVOIR mission concept is one of four Large Mission Concepts studied in preparation for the US 2020 Astrophysics Decadal Survey (Astro2020). The study was initiated in Jan 2016, under the leadership of a Science and Technology Definition Team (STDT) drawn from the community, including representatives of 11 international space agencies. The records and results from 3.5 years of work by the STDT and the LUVOIR engineering team—with critical input and assistance from the broader scientific and technical communities—were submitted to NASA HQ and Astro2020 in 2019, in the form of a Final Report[1] and additional technical material. Fig. 5.10 shows a graphical summary of LUVOIR's key science objectives, which include the technically challenging goal of finding dozens of habitable exoplanet candidates around Sunlike stars and searching them for signs of global biospheres.

The LUVOIR Team developed two distinct observatory concepts: the 15-m LUVOIR-A (Fig. 5.11, left) and the 8-m LUVOIR-B (Fig. 5.11, right). By studying two designs, we gained better understanding of a complex trade space, revealed how science returns scale with different technical choices, and established robustness to uncertainties such as future launch vehicle capabilities and budget constraints. It is important to recognize that LUVOIR-A and -B represent proof-of-concept point designs within a family of UV/optical/NIR observatories, demonstrating feasibility and providing information for the future. LUVOIR's main features are:

---

[1] Available at https://www.luvoirtelescope.org/.

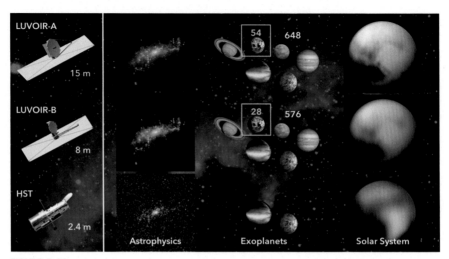

FIGURE 5.10

LUVOIR will revolutionize huge areas of space science. Its sensitivity and spatial resolution open the door to the ultrafaint and ultradistant regime, enabling detailed observations of the full variety of galaxies. LUVOIR dramatically increases the sample size and diversity of exoplanets that can be studied, providing dozens of Earthlike exoplanet candidates that will be probed for signs of life (54 with LUVOIR-A and 28 with LUVOIR-B) and hundreds of nonhabitable exoplanets (648 with LUVOIR-A and 576 with LUVOIR-B). Finally, LUVOIR will provide near-flyby quality observations of solar system bodies.

Credits: NASA/New Horizons/M. Postman (STScI)/A. Roberge (NASA GSFC). HST Pluto image
from Buie et al. (2010).

FIGURE 5.11

(Left) The LUVOIR-A observatory, with a 15-m diameter on-axis primary mirror and four instruments. (Right) The LUVOIR-B observatory, with an 8-m diameter off-axis primary mirror and three instruments. Animations of the telescopes' deployment and pointing may be viewed at https://asd.gsfc.nasa.gov/luvoir/design/.

Credit: A. Jones (NASA GSFC).

- Scalable, serviceable architecture for an observatory at Earth-Sun L2
- A 5-year prime mission, with 10 years of onboard consumables. Nonserviceable components have a 25-year lifetime goal
- Large, segmented, deployable telescopes designed for launch in next-generation heavy lift vehicles with large fairings (e.g., NASA's SLS Block 2, NASA's SLS Block 1B Cargo, Blue Origin's New Glenn, and SpaceX's Starship)

- UV-capable telescopes that are compatible with high-contrast exoplanet observations. Total wavelength range of the studied instrument suite is 100−2500 nm
- A sunshade that is larger but simpler than the JWST sunshield (three layers instead of JWST's five, greatly relaxed requirements on layer positioning after deployment)
- ECLIPS: A high-contrast coronagraph with imaging cameras and integral field spectrographs spanning 200−2000 nm, capable of directly observing a wide range of exoplanets and obtaining spectra of their atmospheres
- HDI: A near-UV to near-IR imager covering 200−2500 nm, diffraction-limited and Nyquist sampled at 500 nm, with high precision astrometry capability
- LUMOS: A far-UV imager and multi-resolution, multiobject spectrograph covering 100−1000 nm, capable of simultaneous observations of up to hundreds of sources
- POLLUX: A high-resolution, point-source UV spectropolarimeter covering 100−400 nm, designed for LUVOIR-A. This instrument study was contributed by a consortium of European institutions, with support from the French Space Agency

4.2 CETUS capabilities to study the cosmic evolution and the origin of life

Summary provided by Sara Heap on behalf of the CETUS team

CETUS, an acronym for **C**osmic **E**volution **T**hrough **U**V **S**urveys (Heap et al., 2019), is a 1.5-m space telescope concept formulated in 2017−19 with support from NASA and industrial and academic partners. The telescope feeds three UV scientific instruments, usually one or two working in parallel. Each instrument has capabilities that are new and unmatched in performance, so the prospect of discovery is high. Fig. 5.12 below shows some examples (top row):

- CETUS′ camera and multiobject spectrograph (MOS) have a wide-field of view, 17.4' × 17.4′ with a resolution of 0.4". It will obtain high-resolution UV imagery for comparison with eROSITA (X-ray), Subaru HSC (optical), Euclid, Roman (near-IR), and ALMA mosaics (mm/submm), and the SKA (21-cm).
- CETUS is sensitive to faint UV sources like the Lyman-a halos around galaxies (shown below) and possibly O VI structures in the CGM.

Other examples, shown in the bottom row:

- CETUS is sensitive in the Lyman UV so it can measure the amount and properties of the warm-hot gas in circumgalactic medium, where arguably the bulk of baryons resides.
- CETUS has a near-UV multiobject spectrograph with a next-generation micro-shutter array similar to the one in Webb but better. The array has over 16,000 slots, each $2.75'' \times 5.5''$ in size. In a single pointing, CETUS will obtain rest-frame far-UV spectra of ~ 70 $z \sim 1$ galaxies or near-UV spectra of 70 star-forming regions in nearby galaxies.
- CETUS is capable responding rapidly to alerts sent from the ground.

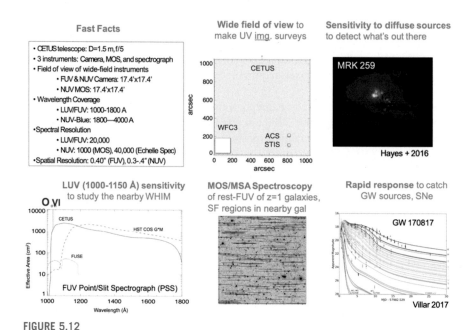

FIGURE 5.12

The new capabilities of CETUS will bring new discoveries.

4.2.1 Focus on star/planet formation and major constituent of planets

Planets form out of gas and dust in debris disks left over from newly formed stars. Dust is both a player and diagnostic of star/planet formation. Dust is a catalyst of star/planet formation by giving a berth for molecular hydrogen to form from two hydrogen atoms. By enabling H_2 to form, it promotes the creation of cold, dense gas, ripe for star-formation. It protects embryonic solar-type stars by shielding them from the harsh radiation of massive stars that have already formed. Dust can also be used as a diagnostic of star/planet formation.

There is an on-going project called PHANGS, which is using ALMA to observe "100,000 star factories" in 74 nearby galaxies. CETUS will contribute to our understanding of star/planet formation in two ways. First, with its wide-field of view, the CETUS far-UV/near-UV cameras will also observe every star-formation region in a PHANGS galaxy at 0.4″ resolution (typically ~20 pc), a finer resolution than ALMA/PHANGS itself. These UV images will give us the "big picture" of the stellar populations and dust structure in the galaxy. The camera filters will not only cover the 1250-4000-Å spectral region with no gaps in spectral coverage, but one rather narrow filter is centered on the 2175-Å extinction bump typical of dust in the Milky Way. Secondly, CETUS will use its wide-field, near-UV MOS to measure the dust extinction curve of every "star factory" observed by ALMA/PHANGS. The spectra will cover 1800–4000 Å at ~20-Å resolution. As shown in Fig. 5.13, previous observations of dust spectra show a tantalizing diversity. NUV spectra of

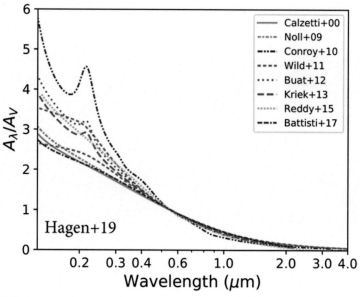

FIGURE 5.13

The diversity of UV dust extinction curves.

100,000 star-forming regions will give insight into the dust properties and surrounding physical conditions that will help us answer questions such as: How does star/planet formation depend on metallicity? On environment? On location in the galaxy? On star-formation rate of galaxy?

4.3 X UV-SCOPE: ultraviolet spectroscopic characterization of planets and their environments

Summary provided by Evgenya Shkolnik on behalf of the UV-SCOPE team

While lower regions of planet atmospheres are being studied at longer wavelengths with HST, and soon with JWST and the ground-based ELTs, UV spectroscopic data are sorely lacking. To fill this need, we are developing a new mission concept for NASA, called UV-SCOPE (UltraViolet Spectroscopic Characterization Of Planets and their Environments). UV-SCOPE will study planetary upperatmospheric formation, composition, temperature, dynamics, and evolution by probing the planet atmospheres via transmission spectroscopy in the UV. Simultaneously, UV-SCOPE will measure the role that host stars play on planetary mass-loss and atmospheric photochemistry, including a planet's potential habitability. Specifically, UV-SCOPE will probe three underexplored planet attributes (Fig. 5.14).

- *The Planet's Exosphere ($\lesssim 1$ nbar):* How much mass is being lost to space? What is the range in mass-loss rates and atomic fractionation across the diverse planet population? We will measure UV transits of planets in search of Ly-a absorption

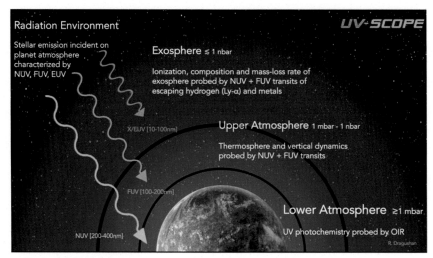

FIGURE 5.14

UV-Scope science sketch.

as well as ionized metals which may be also escaping with the hydrogen, as has been observed with HST transits, e.g., for WASP121b by Sing et al., (2019).

• *The Planet's Upper-Atmosphere (1 mbar − 1 nbar):* What roles do the upper-atmospheric properties (composition, photo-processes, vertical mixing, and temperature structure) play in the escape processes? With low-resolution NUV and FUV spectra, UV-SCOPE will also be sensitive to the upper-atmosphere conditions of hot and ultrahot Jupiters, primarily due to the high opacities of SiO and CO in those regions (Line et al. in prep.)

• *The Radiation Environment and its effects on all levels on the planet atmospheres:* What influence does the high-energy stellar environment have on atmospheric evolution and habitability? For this, we will monitor the stars to assess their UV emission during quiescence, flares, and measure the flare frequencies distributions (e.g., Loyd et al., 2018). In addition, we will use the UV spectra as guides to building new stellar upper-atmosphere models, which will be used to predict the EUV emission and provide full SEDs for irradiating model disks and planets.

By observing both the star and its planet, UV-SCOPE will be able to simultaneously study the *cause and effect* of UV radiation in exoplanets systems.

In order to accomplish these science goals, we are designing UV-SCOPE to be a 60 cm telescope equipped with a two-arm spectrograph: FUV at medium-resolution and NUV at low-resolution with simultaneous wavelength coverage from 120 nm to capture Ly-α out to 400 nm. The targets system will be a wide range of stars (AFGKM), each hosting at least one transiting planet. Planet radii will also span a wide range, from small terrestrial planets to gas giants. We will also include young systems, whose numbers are increasing thanks to dedicated searches to study the evolution of planetary systems.

4.3.1 ESCAPE: exploring the drivers for atmospheric evolution
Summary provided by Kevin France on behalf of the ESCAPE team

Owing to their large number and strong atmospheric impacts, EUV photons are the primary agent for atmospheric escape on planets orbiting cool stars (F, G, K, and M stars). However, previous EUV instruments lacked the sensitivity to survey exoplanet host stars in this critical spectral band. The Extreme-ultraviolet Stellar Characterization for Atmospheric Physics and Evolution (ESCAPE) mission is an astrophysics Small Explorer employing ultraviolet spectroscopy (EUV: 7–82 nm and FUV: 128–160 nm) to explore the high-energy radiation environment in the HZs around nearby stars. ESCAPE provides the first comprehensive study of the stellar EUV environments that directly influence the habitability of rocky exoplanets (see Fig. 5.15). In a 2-year science mission, ESCAPE will provide the essential stellar characterization to identify the star-planet systems most conducive to habitability and provide a roadmap for future life-finder missions (e.g., OST, LUVOIR, and Habex).

Optical and near-infrared photons heat the surface and troposphere. NUV (180–230 nm), FUV (91–180 nm), and X-ray (1–10 nm) photons are absorbed in the middle and upper-atmosphere where they photo-dissociate molecules. EUV photons (10–91 nm) are absorbed high in the atmosphere (i.e., in the thermosphere) where they ionize atoms and molecules. Liberated electrons collisionally heat the surrounding gas, increasing the scale height of the atmosphere and potentially leading to the formation of a hydrodynamic outflow (i.e., rapid atmospheric escape). The stability of Earthlike atmospheres depends critically on the EUV irradiance (Johnstone et al., 2015, 2020, Fig. 5.1). Higher EUV flux from the young Sun (Tu et al., 2015) could have led to $10\times$ greater oxygen loss rates and $90\times$ greater carbon loss rates by increasing the suprathermal or "hot" population of these atoms (Amerstorfer et al., 2017). In highly irradiated planets, the outflow is sufficiently rapid that heavier elements (O and C) can be dragged along through collisions with the lighter hydrogen, as observed on Hot Jupiters (Vidal-Madjar et al., 2004; Linsky et al., 2010; Ballester and Ben-Jaffel 2015). Free electrons produced by stellar EUV photons can attain altitudes much greater than ions, producing an ambipolar electric field that leads to a nonthermal ionospheric outflow (O+ and N+ winds; Dong et al., 2017).

EUV spectra of exoplanet host stars are scarce. The only previous EUV astronomy mission, EUVE (Bowyer, 1991), obtained spectra of ~ 15 cool main sequence stars, including 5 M dwarfs. Previous observations were heavily biased toward the most active stars. The very modest <2 cm^2 effective area of the EUVE spectrometers precluded useful spectroscopic observations of stars with more solar-like activity, except for the nearby alpha Cen system and the F4 subgiant Procyon. No observed EUV spectra exist of the optically inactive M dwarfs (i.e., Ca II H and K equivalent widths < 1.0 Å, e.g., France et al., 2016) that will be optimal for

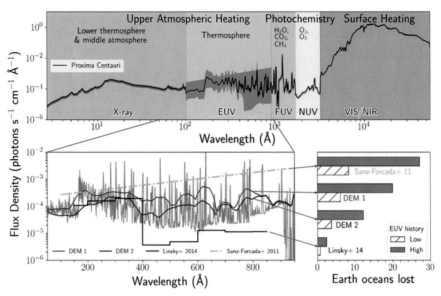

FIGURE 5.15

(top) A reconstructed SED of the nearby planet hosting star Proxima Cen highlights the influences of different parts of the stellar SED on an orbiting planet. The shaded region indicates the relative uncertainty on the intrinsic flux at each wavelength. (bottom left) Different reconstructions of the EUV spectrum of the M dwarf Proxima Cen show factors of 3–100 flux discrepancies: differential emission measure models in red, blue, and gray (Drake et al., 2019); scaling relations based on only FUV in black (Linsky et al., 2014) and only X-rays in orange (Sanz-Forcada et al., 2011). Corresponding hydrogen mass lost from an Earthlike planet (in units of Earth oceans) from 10 Myr to 4.8 Gyr is shown at bottom right for high (shaded gray) and low (hatched) EUV histories of typical M dwarfs (Ribas et al., 2017). ESCAPE resolves the factor of $\sim$30 differences in atmospheric mass-loss rates by directly measuring the EUV spectrum to 40% accuracy (including photometric and systematic uncertainties).

biosignature searches with JWST, 30-m telescopes, and the Origins Space Telescope. The lack of direct EUV data hampers our ability to understand how habitable atmospheres evolve with time, and to design definitive biosignature searches for planets orbiting stars of different masses.

ESCAPE explores the HZ radiation environment around nearby stars with two spectroscopic surveys: SEEN and DEEP. The SEEN survey quantifies the EUV and FUV irradiance from 200 nearby ($d < 80$ pc) F − M stars. ESCAPE observes stars with a range of ages and activity levels, and places an emphasis on stars with known exoplanets. SEEN's broad view is complemented by the DEEP survey: a focused campaign of long duration observations of a select number of stars. Each of the 24 prototypical stellar systems will be observed for 2 weeks to measure their EUV flare frequency distributions and coronal mass ejection (CME) rates.

Harra et al. (2016) reviewed techniques for remote detection of CMEs and determined that coronal dimming, the observational approach employed by ESCAPE, is the only feature consistently associated with CMEs that can be used to detect and quantify these events on other stars.

Large gains in EUV sensitivity (10−100×) over previous missions enables the ESCAPE surveys. ESCAPE employs a grazing-incidence telescope that feeds a EUV and FUV spectrograph, with an instrument design optimized for EUV spectroscopy. The ESCAPE instrument consists of a 46 cm Hettrick-Bowyer telescope (grazing-incidence Gregorian analog; Hettrick and Bowyer, 1984) feeding a four-channel spectrograph. Two channels cover the short-wavelength EUV (7−56 nm) by dividing the telescope beam onto two different grazing incidence (GI) gratings. The other two channels utilize the zero-order reflection off of the GI gratings to cover the complimentary long-wavelength EUV (62−82 nm) and FUV (125−165 nm). Spectral resolution ranges from 1.5 to 6 Å from the EUV to FUV grating modes, and all channels are imaged onto the same MCP detector. The ESCAPE instrument covers these spectral ranges simultaneously with a fixed optical configuration that requires no mechanisms or bandpass limiting optics during normal science operations. ESCAPE would launch in mid/late-2025 and conduct a two-year science mission to identify and understand those star-planet systems that are conducive to the formation of habitable environments.

4.4 The European ultraviolet-visible observatory (EUVO)

Summary provided by Ana I. Gómez de Castro on behalf of the EUVO consortium

EUVO is a concept for a large (6−10 m) size UV-Visible observatory pushed forward by the European astronomical community to address strategic research lines for the coming decades, such as the investigation of the origin of life or the study of the Cosmic Web.

Astrobiology requires very efficient observation of exoplanetary atmospheres that can only be achieved by making use of the strong UV resonance transitions (HI, OI, CI) and the UV molecular absorption bands (O_3, O_2). The evolution of planetary atmospheres depends on the parent star magnetic activity; UV radiation from the star is absorbed by the planetary atmosphere driving planetary winds and atmospheric escape. Any comprehensive study requires simultaneous measurement of the properties of the planetary atmosphere and the stellar activity; this is naturally achieved at UV wavelengths. The stellar UV radiation field also affects the formation of amino acids in comets and the evolution of primitive bacteria in the oceans (see Estrela and Valio, chapter). Habitable planets grow in proto-stellar discs under UV irradiation, a by-product of the accretion process that drives the physical and chemical evolution of discs and young planetary systems. The electronic transitions of the most abundant molecules are pumped by this UV field that is the main oxidizing agent in the disc chemistry and provides unique diagnostics of the planet-forming environment that cannot be accessed from the ground. Knowledge of the variability of the UV radiation field is required for the astrochemical modeling

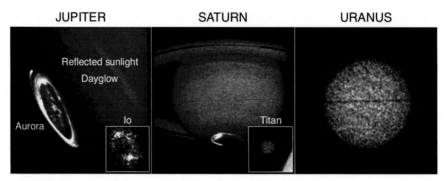

FIGURE 5.16

A collage of some solar system UV targets. From left to right, these include the atmospheres, auroras and airglow of Jupiter, Io, Saturn, Titan, and Uranus.

of protoplanetary discs, to understand the formation of the early planetary atmospheres and the photochemistry of the precursors of life (see Gomez de Castro and Canet, chapter).

The study of the atmospheres and the magnetospheres of Solar System planets provides fundamental clues for the understanding of exoplanets observations, such as the detectability of auroras or weather patterns (see Fig. 5.16).

The cosmic web is the most sensitive tool to study the forces acting on the Universe on the large scale, and during its evolution: gravity (and dark matter), dark energy, radiation and the growth of the magnetic fields that pervade it. Most of the missing baryons are expected to be in the hot intergalactic medium. The thin galactic halos act as transitional layers channeling chemically processed matter from the galaxies into the intergalactic medium (IGM) and feeding the galaxies with fresh material to feed the star-formation cycle. The chemical abundance of the IGM and the galactic halos traces the main feeding routes and requires of the remarkable sensitivity of the UV resonance transitions to detect the gas and measure its metallicity given the low column densities involved.

The main scientific requirements for EUVO are gathered in Table 5.1. Europe's industry and academia is well prepared to support the technological implementation: instrument design, detector technology, optical components, coatings … making feasible the development of a main instrument for LUVOIR, if finally selected as NASA's flagship, or a European led mission: the European UV Observatory (EUVO).

EUVO architecture can be optimized to a large primary (5-m) fitting into an Arianne VI launcher, thus reducing the risks (and costs) of a deployable mirror for optical-UV applications. This, together with the technological improvements since the time when HST was built (more than 30 years ago), will easily result in more than an order of magnitude improvement in sensitivity, meaning a gigantic step forward in astrophysical investigation. EUVO could fit within the cap of an ESA-led L mission developed in collaboration with other space agencies (NASA, JAXA, ROSCOSMOS …).

Table 5.1 Main scientific requirements to EUVO.

Characteristic	Target values	Science topic
L2 or HEO orbit		Variability studies in all topics (AGNs, stellar astrophysics, Solar system research)
Large focal planes	FoV:10×10 arcmin^2 Angular resolution:< 0.01 arcsec	Efficient instrumentation for galactic and extragalactic surveys
UV spectral coverage	90–320 nm	From the Lyman continuum to the Werner H2 bands for ISM, and IGM studies, TTSs disks, planetary aurorae, HI Lyman alpha, CIV, OVI in the IGM/ISM and CO, HD
Integral field spectroscopy	R = 500–3000	Efficient instrumentation to characterize astronomical sources in surveys. Narrow-band imaging in the FUV of extended structures (jets, nebula, star-forming regions, clusters)
Spectroscopy	R = 2000 R = 20,000	Studies of fain compact binaries, cool main sequence stars and brown dwarfs. Young planetary systems. Bulk composition of exoplanets. Detection of photoevaporative flows from exoplanets.
Spectropolarimetry	R = 10,000 (accuracy <0.1%) R = 40,000 (accuracy <0.1%)	Young stars circumstellar environment Stellar magnetic fields
Sensitivity	Increase a factor of 50 with respect to Hubble	Detection of the gas component from transitional to debris disks To study the interplanetary medium in exoplanetary systems To monitor the magnetic activity of M stars within 50 pc To observe Uranus as Jupiter is observed by HST

4.5 Arago

Summary provided by Coralie Neiner on behalf of the Arago consortium

The goal of Arago is to follow the life cycle of matter, and therefore the entire life of stars and planets from their formation from interstellar gas and grains to their death and feedback into the interstellar medium (ISM). During the formation and throughout the life of stars and planets, a few key basic astrophysical properties, especially magnetic fields, stellar winds, rotation, and binarity, influence their dynamics, and thus fundamentally impact their evolution. The associated processes directly affect the internal structure of stars, the dynamics, and the immediate circumstellar environment (as outlined in Fig. 5.17). They consequently drive stellar

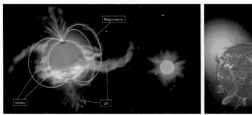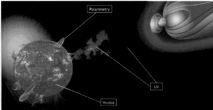

FIGURE 5.17

Left: Sketch of a hot star with its fossil magnetic field lines, channeled polar wind, surface spots, equatorial magnetosphere, corotating interaction regions, and a stellar companion. *Right:* Sketch of a cool star, with its dynamo magnetic field, surface faculaes and plages, wind, a coronal mass ejection, and a bow shock between the star and its planet.

© S. Cnudde.

evolution, but also define the environments of planets, thereby influencing the formation and fate of planets surrounding the stars. Arago will allow us to obtain, for the first time, a full picture of the 3D dynamical environment of stars and their interactions with planets, and explore the conditions for the emergence of life on exoplanets.

Arago consists of a 1.3-meter telescope, equipped with a polarimeter simultaneously feeding two high-resolution spectrographs working in the UV (119–320 nm, R = 25,000) and Visible (355–888 nm, R = 35,000) spectral ranges. This project is prepared by a European consortium led by France as an M class mission candidate for ESA.

The proposed payload consists of a single UV + Visible polarimeter placed near the telescope focal plane. A dichroic behind the polarimeter separates the two wavelength domains to feed two classical echelle spectrometers using cross-dispersion techniques to reach the required spectral resolution. The full spectrum is spread onto two detectors: an MCP for the UV domain and a CCD for the Visible domain. A calibration unit allows us to inject light from calibration lamps in the polarimeter, instead of the stellar light coming from the telescope. The spacecraft and a fine guiding system ensures precise pointing stability (30 mas during 30 min).

Arago will mostly observe stars with magnitude between V = 3 and 10, providing a very high SNR for multi-line averaged spectropolarimetric measurements, and SNR = 10 in chromospheric emission lines of cool (KM) stars. A magnitude-limited legacy survey will be undertaken and immediately made publicly available. Additional statistical surveys, snapshot targets, targets for detailed 3D mapping, and targets of opportunity (e.g., supernovae) will be chosen following open calls for proposals, in an observatory-type mode.

4.6 The MESSIER surveyor

Summary provided by David Valls-Gabaud on behalf of the MESSIER collaboration.

The proposed MESSIER Surveyor is a small-class space mission designed for exploring the last remaining niche in observational parameter space: the very low surface brightness universe. At a 900 km orbit, the mission will drift-scan the entire sky in six bands covering the 200−1000 nm range, reaching the unprecedented surface brightness levels of 34 and 37 mag/arcsec^2 in the optical and UV, respectively. These levels are required to achieve the two main science goals of the mission: to critically test the LCDM paradigm of structure formation through (1) the detection and characterization of ultrafaint dwarf galaxies, which are predicted to be extremely abundant, but remain unobserved; and (2) tracing the cosmic web, which feeds dark matter and baryons into galactic halos, and which may contain the reservoir of missing baryons at low redshifts.

Remarkably, and even though the mission has not been designed for these goals, a large number of additional science cases appear as free by-products of the mission, which is perhaps not surprising given that the sky has never been surveyed to these depths in surface brightness. Hence, additional science cases range from cosmology to stellar physics through extragalactic physics, and include: the galaxy luminosity function, the intracluster and intragroup light, the cosmological UV/optical background, Ly-alpha galaxies at $z = 0.65$, the BAO scale at that redshift using emission-line galaxies, the recalibration of cosmological distance indicators, the actual extension of galaxies, the detection of warm molecular hydrogen at $z = 0.25$, stellar mass-loss envelopes, debris discs, and of course the first reference photometric UV-optical catalog which will have a unique legacy values. The entire astronomical community will benefit from the time-domain full-sky maps that MESSIER will provide.

To achieve these goals, the payload consists of an off-axis TMA telescope with a 50 cm primary mirror at f/2.3 which produces an ultrastable PSF with ultralow wings. No lenses can be used, as not only they create multiple scatterings, but they also produce Cerenkov radiation at levels larger than the faint surface brightness that must be detected. The design includes extreme baffling to ensure a minimal stray-light contamination. The flat focal plane array is divided into 12 independent CCDs, each designed to have a maximum quantum efficiency at the central wavelength of each filter (see Fig. 5.18) and a set of narrowband and broadband filters centered at 200 nm. The design is very robust, as there are no moving parts, and the cooling is passive, ensuring that the power consumption remains low. The CCDs are read in TDI mode to ensure flat field corrections below the 0.0025% level. Each detector is optimized to have QE>80% in each band. With a scale of 1 arcsec per pixel and a FOV of $4° \times 2°$, MESSIER is 200 times better than HST in terms of surface brightness detection.

On the issues related to the origins of life, the MESSIER Science Team has identified the following science cases which will be free by-products of the maps produced by the mission:

[1] The *time-domain mapping of the Zodiacal cloud.* MESSIER will provide a 2 degree-wide multi-band map of the Zodiacal cloud at a solar elongation of 90°, with a time sampling rate of one exposure every 90 min. This will enable the measure of local heterogeneities, asteroidal bands and cometary trails with unprecedented detail. The multi-band maps will allow the inversion of line of

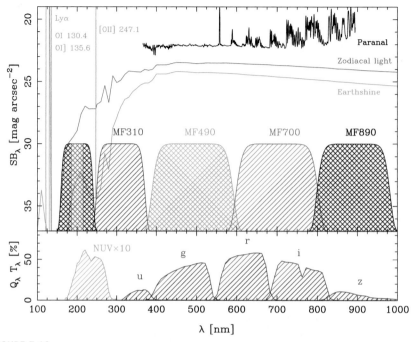

FIGURE 5.18

Top: The surface brightness of the sky at Paranal compared with the emission of the zodiacal light and the earthshine as measured by HST. Not only the strongly variable sky emission (in time and space) is no longer an issue in space, but in addition the gain is over 4 mag/arcsec^2 at red wavelengths, and at 200 nm these two foregrounds become negligible while keeping the throughput at 85%. The proposed set of filters for MESSIER, optimized for characterizing unresolved stellar populations and measuring the cosmic web is also shown. Bottom: The overall throughput of the SDSS, and GALEX's NUV (multiplied by 10). MESSIER is more efficient in SB than any other telescope ever by over two orders of magnitude across the full 200–1000 nm range.

sight measures to constrain the properties of interplanetary dust particles and their growth. This is particularly important given that, at the time of the Late Heavy Bombardment, the Zodiacal cloud was much denser than today, and the small relative velocities of dust particles released by comets in the inner Solar System, combined with their fluffiness, would enable them to survive atmospheric entry and enrich in complex carbonaceous compounds the atmospheres of the early telluric planets. This could also lead to constraints on scenarios based on the panspermia hypothesis.

[2] The *detection of the remnants of the formation of planetary discs.* The study of debris discs is closely coupled with exoplanets and will eventually place our Solar system in its proper context. Planetesimals around other stars are not

directly detectable but their mutual collisions generate enough dust to produce a cross-sectional surface which is large enough to be observable either through thermal emission of large dust grains at midinfrared wavelengths and long-wards, or observable through light scattered by small grains in the optical and UV. Surveys with the ISO, Spitzer, and Herschel satellites have discovered hundreds of dusty debris discs among the thousands of stars in our neighbor-hood (<25 pc). At optical wavelengths, only about two dozen debris discs have been discovered because of their faint brightness and because of the require-ment for high-contrast imaging. The typical sensitivity in the optical is 24 mag/arcsec^2 at 10 arcsec from the star with HST/STIS. MESSIER will offer a full-sky survey of debris discs in the UV/optical with a vastly superior sensi-tivity for discs larger than 5 arcsec in projected radius. The potential of MESSIER to detect debris discs in the MF490 band is simulated in Fig. 5.19 using the most complete catalog of nearby stars and shows that hundreds of

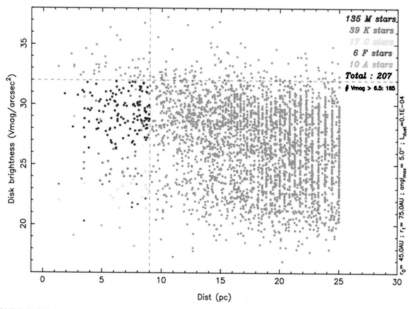

FIGURE 5.19

Potential of MESSIER to probe debris discs around nearby stars. The simulation is based on the Gliese Catalog of nearby stars, and uses a typical fractional dust luminosity of $L_{dust}/L_* = 10^{-5}$, and characteristic inner and outer radii of 45 and 75 au, respectively, for the belt. It assumes also that the PSF subtraction is fully effective only beyond 5 arcsec and, hence, debris discs with angular radii smaller than this limit cannot be recovered (gray-colored stars). Overall, hundreds of disc debris will be measured by MESSIER.

debris discs around a variety of stars of different spectral types will be characterized by MESSIER.

[3] The *detection of the remnants of mass-loss episodes in stars.* The emission from stars is not pointlike, as stars undergoing mass-loss get surrounded by circumstellar shells containing dust particles that scatter light from the central star itself, from nearby stars, or from the interstellar radiation field. Key for the origin of life is the understanding the chemical evolution of the Solar Neighborhood, and this depends on the mass lost by nearby red giants and supergiants, which dominate the mass return budget. Yet the actual mechanisms of mass-loss are not understood. Processes such as turbulent pressure generated by convective motions, combined with radiative pressure on molecular lines, might initiate mass-loss which coupled with Alfven waves generated by magnetic fields drive the episodes of mass-loss. Pulsations and radiation pressure on newly formed dust grains may take over in AGB stars, but the detailed understanding of these mechanisms is lacking. Similarly, the interaction with the surrounding interstellar medium creates shocks in the wind-ISM interface, which have only been serendipitously detected recently. Long tails such as the one detected around Mira (2.5° long) by GALEX, and shells around Betelgeuse and AGB stars have led to the concept of astrospheres. In the UV domain, the FUV emission is inferred to be H_2 line emission collisionally excited by hot electrons in shocked gas, but much remains to be studied. By reaching unprecedented surface brightness levels, MESSIER will systematically detect all astrospheres larger than 5 arcsec in projected radius, providing UV-optical detailed maps to characterize their properties, bringing unique constraints on the actual chemical evolution on solar and nonsolar type stars.

4.7 The cosmological advanced survey telescope for optical and UV research (CASTOR)

Summary provided by Jason Rowe and Patrick Côté on behalf of the CASTOR consortium

Cosmological advanced survey telescope for optical and UV research (CASTOR) is a wide-field, nearly diffraction-limited space telescope concept that has been the subject of recent study and technology development activity sponsored by the Canadian Space Agency (CSA). Recently identified as a top priority in Canadian space astronomy for the 2020s, the project is aiming toward a launch around 2027. In addition to Canada, prospective partners in the mission include India, JPL and the UK.

The 1m CASTOR telescope will produce panoramic imaging of the UV/optical (150–550 nm) sky using a three mirror anastigmat (TMA) design that provides HST-like image quality (FWHM = 0.15″) over a wide-field of view (0.25 deg^2) in three filters, simultaneously. CASTOR, which is currently baselined to a (minimum) five-year lifetime, is being designed for launch on the Indian PSLV rocket, which will place the telescope into a sun-synchronous, low-earth orbit at an altitude of

800 km. Although CASTOR will be optimized for wide-field imaging surveys, it will also feature low-resolution spectroscopic capabilities over the 150–400 nm range using a retractable grism. A precision photometric capability in each of its three channels will be possible optimized detectors in each of its three focal planes and an optical design that defocuses the light from bright targets. Finally, a medium-resolution multiobject spectroscopic capability in the 150–300 nm region may be provided by a DMD spectrograph.

To summarize, the CASTOR suite instrumentation is designed to provide:

- High-resolution (FWHM = 0.15″) and wide-field (0.25 $\deg^2$) imaging in three broadband filters, simultaneously: UV (150–300 nm), u (300–400 nm) and g (400–550 nm).
- Low-resolution, slitless spectroscopy in the UV (150–300 nm) and u (300–400 nm) channels covering the full 0.25 $\deg^2$ FOV. A single grism mechanism provides R ∼ 300 and ∼420 in the UV and u channels, respectively, for $\Delta\lambda$ ∼ 2px.
- Multi-slit, medium-resolution (R = 1,000–2,000) UV spectroscopy (150–300 nm) in a parallel field measuring 210″ × 120″ and offset by 3′-4′ from the edge of the imaging field. Up to ∼600 slits will be possible with a 4 × 2K digital micromirror device (DMD).
- Precision photometric monitoring (10 ppm) in the UV-, u- and g-bands for bright stars using three optimized CMOS detectors equipped with transmissive diffuser plates.

The CASTOR science program will be implemented through a combination of guest observer (GO) programs and legacy surveys that span a range of science topics. A major component of the science plan will include a 1.8-yr "primary survey" that will image an area of ∼7700 $\deg^2$ (defined as the overlap region between the LSST Wide-Fast-Deep, the Euclid Wide, and the Roman High Latitude surveys) to a u-band depth roughly 1.3 mags fainter than the final (10-year) depth of LSST.

Over last decade large imaging and radial velocity surveys have revealed that exoplanetary systems are common. These surveys have also shown that physical properties of exoplanets, such as mass, radius and orbit show a wide range of diversity, including the discovery of new classes of exoplanets, such as *Super Earths* and *mini-Neptunes* that do not have direct solar system analogs. Understanding how exoplanetary systems form and evolve is of immense interest as it directly leads to insight into how potential life-hosting *Earthlike* planets could form. The CASTOR mission provides a unique opportunity to leverage its large aperture to obtain ultra-precision photometry and low-resolution spectroscopy in the UV and blue range of the electromagnetic spectrum. The CASTOR bandpasses, shown in Fig. 5.20, are unique relative to any operational astronomical spacecraft, except the heavily oversubscribed and aging Hubble Space Telescope. The CASTOR science team has identified a number of key science questions that will be addressed through GO and legacy surveys, including:

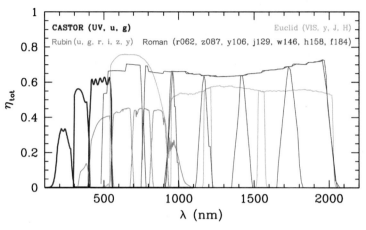

FIGURE 5.20

Passbands for CASTOR (blue) compared to those of the Euclid (orange), Roman (red), and Rubin (green) telescopes, illustrating the CASTOR's uniqueness and synergy with these next-generation facilities.

1. **Transit spectroscopy.** Transit surveys ($\sim$10 ppm precision) of $\sim$50 bright transiting exoplanet systems will be used to measure atmospheric scale heights and scattering properties. These observations will help us understand how atmospheres survive and evolve and are essential for the interpretation of optical and IR measurements as UV-blue observations unique sample the Rayleigh slope to establish the present of scattering hazes and cloud decks in the upper atmospheres of exoplanets.

2. **Exoplanet albedos.** The large aperture of CASTOR enables ultraprecise ($\sim$1 ppm) phase curve measurements for exoplanets around relatively bright stars. Planets ranging from Jupiter to Earth-sized should be accessible. Phase-curves provide unique measurements of scattering angles to reconstruct particle size distributions, cloud properties and potentially even meteorological studies.

3. **Kepler HZ planets.** Follow-up multiwavelength transit measurements for candidate Earth-sized exoplanets in the habitable zones of Sunlike stars will enable candidate validation and elimination of false-alarms from the Kepler HZ sample. CASTOR will provide a true eta-Earth measurements to enable validation.

4. **Exoplanets in Globular Clusters and the Bulge.** In its wide-field imaging mode, transit surveys in the UV, u, and g-bands targeting dense star fields (i.e., bulge fields or globular clusters) would provide a window into the number and properties of exoplanets in new or underexplored environments and metallicity regimes.

5. **Bulk composition of rocky material in white dwarfs.** UV/blue-optical imaging and low-resolution spectroscopy from CASTOR will make it possible to efficiently identify and derive abundances (Mg, Fe, and Ca) for thousands of

metal-polluted white dwarfs. Such systems are believed to have accreted heavy materials from planetesimals, asteroids, small planets and comets.

4.8 SIRIUS: Stellar and ISM research via in-orbit ultraviolet spectroscopy

Summary provided by Martin A. Barstow on behalf of the SIRIUS team

SIRIUS is a proposed high-resolution EUV spectroscopy mission to perform observations of nearby stars and the ISM. The main goal of the SIRIUS mission is to understand this cosmic feedback system within the local bubble and how it ultimately influences the habitability of our region of the Galaxy. The payload is based on novel normal-incidence diffraction grating technology, which provides high effective area and spectral resolution, compared to grazing-incidence systems, allowing **observatory-class science to be delivered in a low-cost package**. The instrument has already been proven in two suborbital space flights and is ready to be deployed on a satellite platform with minimal development.

4.8.1 Scientific objectives

The formation and evolution of stars, their interaction with interstellar material, and the ultimate effect of all the various physical processes on their planetary systems are key but poorly understood facets of Galactic evolution. Crucial elements concern the levels of activity in main sequence stars and the resulting stellar winds, which can directly affect planetary environments on a range of time-scales. In addition, stellar winds control the flow of material and flux of cosmic rays from the Galactic environment, which also have a potential influence on planetary climate. Ultimately, stars recycle material back into the ISM, enriching Galactic metal content, through mass-loss processes associated with the production of white dwarfs and supernovae. All the important processes involved in these stellar life-cycles are traced by the presence of hot (10^5–10^7 K) gas, which can be efficiently investigated in the EUV range. SIRIUS will perform unique EUV spectroscopy to diagnose the density, temperature, composition, structure, and dynamics of these hot astrophysical plasmas.

The program addresses a number of science themes, and concentrates on particular subthemes, as outlined below:

> ***What are the conditions for planet formation and the emergence of life?***
> *Life and habitability in the Solar System (effects of stellar activity on habitability)*
> ***How does the Solar System Work?***
> *From the Sun to the edge of the Solar System (interaction of the heliosphere and ISM)*
> ***What are the fundamental physical laws of the Universe?***
> *Matter under extreme conditions (degenerate matter, accretion onto compact objects)*

How did the Universe originate and what is it made of?
The evolving violent Universe (lifecycles of matter and the Galactic environment)

4.8.2 Science requirements

EUV spectroscopy provides key, yet not fully explored, diagnostics of the physical conditions in the many locations where hot gas is found. However, in the past, observers have only been able to examine the bulk of material, without being able to separate out the distinct gas components present due to limited instrument spectral resolution and throughput. Proven new developments in normal-incidence/multi-layer grating and detector technology now allow high effective area and relatively spectral resolution $(R \sim 5,000)$, providing the extradimension of radial velocity measurements, a powerful tool to separate multiple spectral emission or absorption components of hot plasma diagnostic lines (Table 5.2).

Table 5.2 The primary scientific objectives for the SIRIUS mission.

Primary scientific objectives	Targets	Scientific products
1) Examine the structure of stellar coronae	19 main sequence stars and giants	Shape of spectral line wings; constraints on dynamo and heating models
2) Determine effects of coronae on planets andand vice-versa	10 planet hosting stars	EUV radiation levels and planetary irradiance
3) Observations of young stars	9 stars belonging to young associations	Distinguish between coronal emission and an accretion stream
4) Investigate abundance anomalies	19 main sequence stars and giants - some overlap with 1), some different targets	Relative abundance of elements with low and high first ionization potentials
5) Monitoring observations of flaring stars	16 G-M stars	Measure flows and line widths in flares
6) Study the evolution of white dwarfs	40 white dwarfs	Measure abundances for hot white dwarfs; determine radiative levitation/diffusion balance and effect on atmosphere
7) Probe the structure of local interstellar gas	40 white dwarfs (same observations as for white dwarf evolution)	Determine He ionization fraction; measure He I abundance
8) Examine hot plasma effects	4 flare stars	Doppler imaging

4.8.3 Mission concept

Our proposed SIRIUS concept is based on the J-PEX, sounding rocket-borne near-normal-incidence spectrometer (Cruddace et al., 2002; Kowalski et al., 2006, 2011). This will be optimized to cover the spectral range ∼170–260Å in two discrete sections, each delivering simultaneous high resolving power (∼5,000) and effective area (∼10 cm²). These wavelengths encompass critical spectral features, allowing full application of the plasma diagnostic techniques already used successfully in solar research to a sample of stars within ∼100 pc of the Sun. High-resolution spectroscopy of hot plasmas will allow unambiguous detection and measurement of weak emission lines and absorption features, and the study of source structure and dynamics through measurement of line profiles and Doppler shifts. The mission will require a three-axis stabilized spacecraft bus with a pointing accuracy of 60 arcsec. To achieve the required spectral resolution, knowledge of the attitude, from a coaligned visible light telescope, must be ∼1 arcsec, with any drift below 1 arcsec per time sample. The spacecraft will allow acquisition of targets within 60 degrees of the plane orthogonal to the Earth-Sun line for efficient scheduling of observations and access to all possible targets within a 6 months period. SIRIUS will be placed into L2 orbit, which allows high observing efficiency and confers a low Geocoronal background advantage compared to low-Earth orbit. A baseline mission lifetime of 2 years will allow observation of approximately 100 individual targets, many will be studied for time variable behavior and/or monitored on repeated visits, with integration times in the approximate range 10–1,000 ks (Table 5.3).

Table 5.3 Mission summary table.

Category	Requirements
Launch Vehicle	Shared launch on Ariane 62
Trajectory design	Science orbit: L2
Flight system	Design lifetime: 5 years, nominal mission 2 years Consumables: 5 years Payload: Single EUV spectrograph 　　Mass: 60 kg, Power: 20 W 　　Observing Mode: 3-Axis Platform 　　Pointing accuracy: 1 arcmin 　　Pointing stability: 1 arcsec/s Average science data acquisition rate: 20 kb/s
Spacecraft bus	Example: High TRL design based on Airbus Hub for Falcon 9 or TAS bespoke design
Data return Strategy	Downlink volume <1.9 Gb per day, X-band
Telecommunications	TBD by ESA
Ground systems	1 contact per day via ESTRACK ground station
Solar array	1.9 m² triple junction GaAs cells
Propulsion (for attitude control and station keeping)	Hydrazine monopropellant

4.8.4 Mission heritage and technology development

The mission concept has been proven in a successful NASA-supported sounding rocket program, in a US–UK collaboration. Two flights (Feb 2001 and Oct 2008) yielded high-resolution EUV spectra of the hot white dwarfs G191-B2B and Feige 24, demonstrating the unique capability of detecting the absorption complex of ionized interstellar helium and numerous narrow weak metal lines. The key instrument technologies have been developed in Europe. Normal-incidence gratings were developed originally for J-PEX at Zeiss International, with similar technologies supplied by HORIBA–Jobin Yvon for several solar instruments in the EUV. High-resolution imaging MCP detectors, for recording the spectrum, have been developed by both Universities of Leicester and Tübingen, with strong satellite heritage (ROSAT) and the key MCP components supplied by Photonis (Brive, France). Alternative, emerging CCD and CMOS technologies, with multiple mission heritage, are available from Teledyne e2v (UK). Consequently, the proposed mission is at a mature phase of development and carries minimal risk in its implementation, allowing a relatively rapid and easily managed C/D phase for launch readiness in 2027.

4.9 A cubesat for astronomical spectroscopy

Summary provided by Noah Brosch on behalf of the team

The UV is the spectral domain where the sky background (at orbital night) is the faintest, some two orders of magnitude fainter than the sky background at the best ground-based astronomical sites. This implies that very deep observations can be obtained using relatively modest optics and extended observing times on a dedicated platform. This part of the spectrum cannot be observed from the ground because of the atmosphere, primarily the ozone absorption but also the significant scattering of the photons.

The concept presented here is for a nanosatellite based on the cubesat philosophy that would perform a very specific task having to do with the nature of the interstellar medium, but which could be used for other astronomical investigation as well.

The instrument to be orbited consists of a small imaging telescope for the UV equipped with either an objective grating (OG) or prism (P) could provide spectroscopic observations of all the sources in its FOV. The design is based on a small imaging telescope designed at the IIA.

The imaging optics following the OG forms a (linear) spectrum of every source whose light enters the optical system. In contrast with this, a P forms a single spectrum for each source, but the dispersion is not linear because of the changing index of refraction with wavelength. The focal plane is equipped with a UV-sensitive detector, possibly an intensified sCMOS followed by the electronics chain (see Fig. 5.21).

The main project that could be attempted with such a nanosat would deal with the measurement of the equivalent width of an interesting feature called "the 2,174Å interstellar absorption band." This feature appears in the spectra of virtually all stars

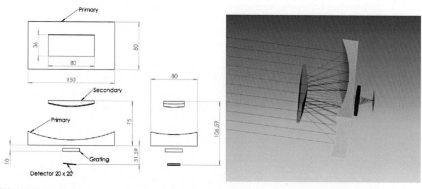

FIGURE 5.21

Left: Schematic description of a small reflecting telescope for the UV equipped with a transmission grating. Right: 3D rendering of a planned IIA UV telescope.

in the Milky Way galaxy and in the stars of other galaxies. The feature was discovered in mid-1960s (e.g., Seaton, 1979), but its origin is not yet well understood. One possibility, supported by the study of Blasberger (2017), is that the feature is produced by UV radiation absorption in Polycyclic Aromatic Hydrocarbons (PAHs, also linked to the "origin of life" question), with the energy partly reemitted in the near-infrared emission on the PAHs. However, until now no systematic survey for the presence and equivalent depth of the 2,174Å in the spectra of stars has been conducted resulting in a small sample of stars in which the feature was measured. The described nanosat, which could be realized in a 4u format, could provide measurements of the band for some 5000 stars.

Additional science that could be performed with such a nanosat could include bright transient source early UV spectroscopy, monitoring of bright variable sources such as supermassive black holes at centers of galaxies, certain aspects of star-formation in galaxies, etc.

4.10 CASSTOR

Submitted by Coralie Neiner on behalf on the CASSTOR team

CASSTOR is a 12U nanosatellite project meant both as a technological demonstrator and scientific demonstrator for UV high-resolution spectropolarimetry on a large wavelength domain. It consists of a 12-cm telescope, a polarimeter, and a spectrograph covering the 135−291 nm wavelength domain (see the outline in Fig. 5.22).

The goals of CASSTOR are:

(1) to demonstrate that the proposed UV polarimeter technology can stand space conditions. The polarimeter is made of a stack of thin birefringent plates of MgF2 in molecular adhesion followed by a MgF2 Wollaston prism. The stack of thin plates can be rotated to perform temporal modulation of the

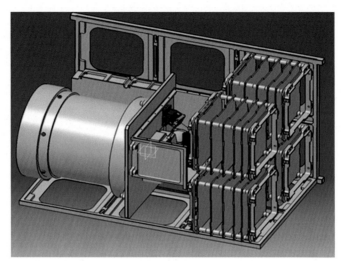

FIGURE 5.22

Model of the CASSTOR nanosatellite showing the telescope (left), payload (middle), and platform elements (right), inside the 12U frame.

polarization, while the Wollaston prism separates the two orthogonally polarized beams. Such a polarimeter is the current baseline for many space mission projects for UV spectropolarimetry, such as Arago, POLLUX on board LUVOIR, and PolStar (see earlier). It is therefore important to increase the technological readiness level (TRL) of this subsystem.

(2) demonstrate that science with high-resolution UV spectropolarimetry on a wide wavelength domain is feasible by producing the very first stellar UV spectropolarimetric measurements. To this aim, CASSTOR will observe very bright stars. In particular, it will study the magnetic field in the circumstellar environment of Ap/Bp stars, the environment of Wolf Rayet stars, the circumstellar disks of Be stars including gamma Cas, and the inhomogeneity of the wind in blue supergiants. No UV spectropolarimetric data of these objects exist as of today. CASSTOR will provide their very first stellar UV spectropolarimetric measurements.

CASSTOR is currently under study by a French consortium with the support of CNES (the French space agency). Its launch is planned in 2025.

4.11 Ultraviolet researcher for the investigation of the emergence of life (URIEL)

Submitted by Ana I. Gómez de Castro on behalf of the team

Life is plenty in planet Earth but whether this phenomenon is ubiquitous in the Galaxy and sustainable over time-scales comparable to stellar evolution is a big unknown. Amino acids are the building blocks of proteins, the biomolecules that provide cellular structure and function in all living organisms on Earth. Laboratory experiments have demonstrated that amino acids grow efficiently in interstellar ices submitted to ultraviolet irradiation (Muñoz-Caro et al., 2002). Also, nucleic acid precursors have been created in the lab starting with just hydrogen cyanide, hydrogen sulfide, and ultraviolet light (Powner et al., 2009). The evidence suggests that first life appeared on Earth earlier than 3.7 Gyr ago (Ga) (Nutman et al., 2016) during a period of heavy meteoric bombardment. Therefore, the contribution of space generated nucleobases to the Earth budget cannot be neglected and competes well with hydrothermal vents in the polymerization of nucleotides and the emergence of long RNA chains (Pearce et al., 2017).

As today, only glycine has been detected in Solar System small bodies: in the Murchison meteorite (Engel and Macko 1997; Kvenvolden et al., 1970), in the samples returned from space probes such as STARDUST (Elsila et al., 2009) and from in-orbit measurements of the comet 67P/Churyumov–Gerasimenko by the Rosetta mission (Altwegg et al., 2016). The detection of other amino acids remains evasive.

Direct measurements of amino acids abundance in comets are hampered by contamination problems and the enormous resources required to reach their surface. Moreover, the investigation is restricted to just a few Solar System bodies and cannot be extended to other planetary systems. Laboratory experiments show that after glycine, the most abundant amino acid is expected to be alanine, with a relative abundance to glycine of $\sim 20\%$ (Muñoz-Caro et al., 2002). Alanine is a chiral molecule, as a result, abundance imbalance between the two mirror images of chiral molecules, also denoted as enantiomers or optical isomers, produces a variation of the incoming circularly polarized radiation at wavelengths where structural differences are the largest, 180 nm for alanine. The strength and properties of the polarization signal depends on the relative concentration of the enantiomers along the line of sight.

The URIEL mission is thought as a pathfinder to evaluate the feasibility of the remote detection of alanine in Solar System bodies by its optical activity at 180 nm. If proven feasible, this technology will enable addressing the source of the chirality imbalance in Earth's biomolecules.

Moreover, URIEL will investigate the distribution and properties of large molecules and small particles at the time of planetary buildup. Hundreds of planetary systems are being formed within 140 parsec around the Sun however, little is known about the processes involved in the disk dissipation, planet formation and the generation of swarms of comets. Spectropolarimetry is a unique tool for this study since exploits at full the information carried by the electromagnetic waves. The unpolarized stellar radiation is scattered by the large molecules and dust particles in the circumstellar environment and becomes polarized. The degree and direction of polarization of this light depends on the microphysical properties of the scatterers (composition, shape, size distribution), the wavelength, and the scattering angle. Indeed, the state of polarization of the scattered light has been shown to be much more sensitive to particle properties than the intensity of the scattered light. These particle properties cannot be derived from intensity measurements. URIEL is designed to monitor the spectropolarimetric variations of young planetary systems in a broad spectral range, from 150 to 600 nm and during two complete stellar rotation periods.

URIEL's main scientific objectives are summarized in Table 5.4.

4.11.1 Mission concept

URIEL is conceived as 50 cm primary telescope with a Ritchey-Chrétien design. The telescope is equipped with a single instrument to obtain spectropolarimetric observations in the 150—600 nm wavelength range in a single exposure. The dispersion

Table 5.4 Scientific requirements - traceability.

Scientific objectives	Scientific requirements	Measurements
Search for signatures of alanine in comets	High sensitivity spectropolarimetry in the 150—250 nm spectral range. Long slit	Rotation of the polarization vector
Mineralogy of solar system and interstellar minor bodies	High sensitivity spectropolarimetry in the 150—600 nm spectral range	Composition
Measure the abundance of small dust grain in the disk atmosphere	High sensitivity spectropolarimetry in the 150—600 nm spectral range	Variation of polarization with wavelength. Variation of the polarization strength with the age of the system
Measure abundance of holes and rings in the discs	High sensitivity spectropolarimetry in the 150—600 nm spectral range	Variation of the polarization strength between systems with similar ages and stellar activity
Measure the size of the stellar magnetosphere	High sensitivity spectroscopy in the 150—350 nm spectral range	Lines and continuum flux
Detect dust clouds close to the star	Monitoring during 1.5—2 stellar rotation periods of the spectropolarimetric signal in the 150—600 nm range	Variations in the polarization with rotational phase.
UV radiation input on the disk	Monitoring during 1.5—2 stellar rotation periods of the spectropolarimetric signal in the 150—350 nm range. Photon counting detector	UV radiation field and variability time-scales.
UV radiation field for ARIEL targets	UV spectroscopy in the 150—600 nm range to overlap with ARIEL photometric bands	UV spectra and flares (if required)

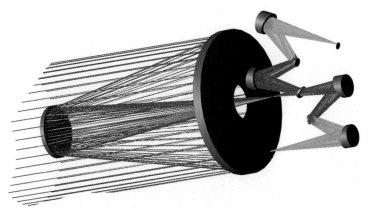

FIGURE 5.23

URIEL telescope concept. The Ritchey-Chrétien mounting telescopes serves to spectropolarimeters based in the known Offner relay concept in which a grating is written onto the surface of the convex mirror present in both systems. The retrieval of the polarization states is accomplished by a polarization module not shown in the images above.

is 600 and enables resolving the spectral features while guarantees enough flux per resolution element for the Stokes parameters to be measured accurately in the full range.

URIEL is designed as a mission with two major sets targets: comets and young planetary systems (T Tauri and Ae/Be Herbig stars). The L2 orbit will optimize the observation schedule and according to our estimates the full program could be completed in three years. The mission will require a 3-axis stabilized spacecraft and a pointing stability better than 1 arcsec.

The telescope concept is outlined in Fig. 5.23. The telescope serves a pair of spectropolarimeters. One of the instruments is meant to cover the range 150−350 nm while the other operates in the 300−600 nm. There is a slit located at the image plane of the main telescope with a spatial extent equivalent to $\pm 10.0''$ in image angular space. After the slit plane a splitting of the beam into the two operating spectral ranges is accomplished by means of a dichroic beam splitter made of either Magnesium or Lithium Fluoride.

5. Summary

The projects described in this contribution are summarized in Table 5.5a−c. Also, some information is added on those projects that did not contribute to the main text. The most noticeable being the 2-m primary space telescope being built by the Chinese Space Agency in support to the China Space Station.

Table 5.5a Summary on UV missions (flying or under construction).

Project	Operator	Main characteristics	Status
HST	NASA and ESA	Observatory-type mission includes instrumentation for imaging and spectroscopy in the 100–320 nm range. Diameter of the primary: 240 cm Orbit: LEO	Flight
XMM-Newton/ Optical Monitor (OM)	ESA	X-ray observatory includes an instrument, OM, with capabilities for UV imaging from 170 to 320 nm. Diameter of the primary: 30 cm Orbit: HEO	Flight
SWIFT/ Ultraviolet/ Optical Telescope (UVOT)	NASA/PSU	γ-ray surveyor mission equipped with an instrument—UVOT—to detect the UV-optical afterglow 170–650 nm. Diameter of the primary: 30 cm Orbit: LEO	Flight
ASTROSAT-UVIT	ISRO	UV observatory operating in the 130–300 nm range with imaging and slitless spectroscopic capabilities. Diameter of the primary: 37.5 cm Orbit: LEO	Flight
Spektr-UF/ WSO-UV	ROSCOSMOS	UV observatory operating in the 115–315 nm range with instrumentation for imaging and spectroscopy. Diameter of the primary: 170 cm Orbit: HEO	Launch scheduled 2025
China space Station telescope	CNSA	UV-optical surveyor with FoV = $1.1 \times 1.1 \text{deg}^2$ and Ang. Res. = 0.15″. NUV filter covers the 255–320 nm range. Diameter of the primary: 200 cm Orbit: LEO	Under construction
CUTE	NASA	Dedicated cubesat mission for exoplanetary research obtained spectra in the 255–330 nm range Primary: 20×8.5 cm Orbit: LEO/Sun-synchronous	Launch scheduled 2021
SPARCS	NASA	Dedicated cubesat mission for exoplanetary research equipped with filters in the 115 –320nm. Diameter of the primary: 9 cm Orbit: LEO/Sun-synchronous	Under construction (launch Fall, 2021)

Table 5.5b Summary on UV projects (list may not be complete).

Project	Teams	Main characteristics	Status
LUVOIR	NASA, EU, JAXA (and TBD)	UV-optical-infrared observatory-type mission includes instrumentation for imaging, spectroscopy and spectropolarimetry in the UV range. Diameter of the primary: LUVOIR-A (15 m), B(8 m) Orbit: L2	Study of concept
CETUS	NASA	UV observatory-type mission with instrumentation for imaging and spectroscopy in the 100−320 nm range. Diameter of the primary: 150 cm Orbit: L2	Study of concept
ESCAPE	NASA	Dedicated mission to EUV spectroscopy (7−165 nm) of nearby stars to measure the radiation driving at atmospheric escape. Diameter of the primary: 46 cm Orbit: LEO	Phase A
EUVO	EU	UV observatory-type mission with instrumentation for imaging and spectroscopy in the 100−320 nm range. Diameter of the primary: 5m to LUVOIR Orbit: L2	Under study
SIRIUS	UK, EU	EUV mission for spectroscopy in the 17−26 nm range with a predefined list of targets. Effective area: 10 cm^2 Orbit: L2	3xsuborbital flights. Study of concept
INSIST	Indian Institute of astrophysics	UV observatory equipped for imaging and multiobject spectroscopy in the 150−550 nm range. Diameter of the primary: 100 cm Orbit: LEO	Study of concept
CASTOR	Nat. Research and Engineering Council of Canada	UV-optical surveyor. Simultaneous imaging coverage three passbands that span the 150−550 nm range. Diameter of the primary: 100 cm Orbit: LEO	Study of concept

Continued

Table 5.5b Summary on UV projects (list may not be complete).—*cont'd*

Project	Teams	Main characteristics	Status
ARAGO	Observatoire Paris-Meudom, EU	UV observatory equipped for high dispersion spectropolarimetry in the 119—320 nm range. Diameter of the primary: 130 cm Orbit: L2	Study of concept
MESSIER	Observatoire de Paris, EU	Imaging surveyor operating simultaneously in 6 bands within the 200—1000 nm spectral range. Diameter of the primary: 130 cm Orbit: L2	Study of concept
URIEL	Universidad Complutense de Madrid, EU	UV-optical observatory for spectropolarimetry in the 150—350 nm range Diameter of the primary: 50 cm Orbit: L2	Study of concept
CAS	Tel-Aviv University, India, Spain	Dedicated cubesat mission for interstellar medium studies (UV bump). Surveyor, suitable for early alarms. Primary: 15 × 8 cm Orbit: LEO	Study of concept
CASSTOR	CNES	Demonstrator cubesat mission for UV spectropolarimetry. Spectral range: 135—291 nm Primary: 12 cm Orbit: LEO	Under study

Table 5.5c Summary on UV LUNAR projects (list may not be complete).

Project	Teams	Main characteristics	Status
Lunar observatory	International Lunar Obs (USA)	UV interferometer to resolve the surface of stars Diameter of the primary: 50 cm Orbit: On ground (Lunar south pole)	Under study
EarthASAP	Universidad Complutense de Madrid, Russia, Japan	UV cubesat mission for monitoring (imaging) in a set of bands in the 115—175 nmn range Diameter of the primary: 50 cm Orbit: Lunar polar	Study of concept

This contribution shows the vibrant scientific and technological environment around the UV missions and space projects. It also shows that the future UV capabilities for astrobiological research are being defined now. The scientific community driving current astrobiological research is deeply involved in them.

Acknowledgments

Prof. Ana I Gómez de Castro wants to acknowledge the work the Network for Ultraviolet of Astronomy and the Working Group for UV Astronomy in gathering some key information included in this contribution. This work has been partly funded by the Ministry of Science and Innovation of Spain through grant: ESP2017.87813-R.

References

Altwegg, K., Balsiger, H., Bar-Nun, A., et al., 2016. Sci. Adv. 2 (5), e1600285.
Amerstorfer, U.V., Gröller, H., Lichtenegger, H., et al., 2017. JGR 122, 1321.
Ballester, G.E., Ben-Jaffel, L., 2015. ApJ 804, 116.
Blasberger, et al., 2017. ApJ 836, 173.
Bourrier, V., Lecavelier des Etangs, A., Dupuy, H., et al., 2013. A&A 551, 63.
Bowyer, S., 1991. ARAA 29, 59.
Brosch, N., 1998. New horizons from multi-wavelength sky surveys. In: McLean, B.J., Golombek, D., Hayes, J.J.E., Payne, H.E. (Eds.). Springer, p. 57.
Buie, M.W., Grundy, W.M., Young, E.F., et al., 2010. AJ 39, 1128.
Byram, E.T., Chubb, T.A., Friedman, H., et al., 1957. Astronom. J. 62, 9.
Cruddace, R.G., et al., 2002. ApJL 565, 47.
Dong, C., Lingam, M., Ma, Y., Cohen, O., 2017. ApJL 837, L26.
Drake, J.J., Kashyap, V.L., Wargelin, B.J., et al., 2019. arXiv 1901, 05525.
Dressing, C.D., Charbonneau, D., 2015. ApJ 807, 45.
Egan, A., Fleming, B.T., France, K., et al., 2018. Proc. SPIE 10699, 106990C.
Elsila, J.E., Glavin, D.P., Dworkin, J.P., 2009. Met. Planet. Sci. 44, 1323.
Engel, M., Macko, S.A., 1997. Nature 389, 265.
Fleming, B., France, K., Nell, N., et al., 2018. JATIS 4, 014004.
France, K., Loyd, R.O.P., Youngblood, A., et al., 2016. ApJ 820, 89.
France, K., Fleming, B., Drake, J., et al., 2019. SPIE 11118, 8.
Gómez de Castro, A.I., López-Santiago, J., López-Martínez, F., et al., 2015. MNRAS 449, 3867.
Harra, L.K., Schrijver, C.J., Janvier, et al., 2016. Sol. Phys. 291, 1761.
Heap, S., et al., 2019. CETUS final Report. arXiv 1909, 10437.
Hettrick, M., Bowyer, S., 1984. NASA Conf. Publ. 2349, 529.
Jewell, H., Jones, et al., 2018. Proc. SPIE 10709.
Johnstone, C., Gudel, M., Stokl, A., 2015. ApJ 815, 12.
Johnstone, C., Bartel, M., Gudel, M., 2020. A&A (in press).
Kvenvolden, K.A., Lawless, J., Pering, K., et al., 1970. Nature 228, 923.
Kopparapu, R.K., 2013. ApJ 767, L8.

Kowalski, M.P., et al., 2006. SPIE proc. 6266, 24.

Kowalski, M.P., et al., 2011. ApJ 730, 115.

Linsky, J.L., Yang, H., France, K., et al., 2010. ApJ 717, 1291.

Linsky, J.L., Fontenla, J., France, K., 2014. ApJ 780, 61.

Llama, A., Shkolnik, P., 2015. ApJ 802, 41.

Loyd, R.O.P., Shkolnik, E.L., Schneider, A.C., 2018. ApJ 867, 70.

Luger, R., Barnes, R., 2015. Astrobiology 15, 119.

Meadows, V., 2017. Astrobiology 17, 1022.

Muñoz Caro, G.M., Meierhenrich, U.J., Schutte, W.A., et al., 2002. Nature 416, 403.

Nikzad, S., Hoenk, M.E., Greer, F., et al., 2012. Appl. Optic. 51, 365.

Nutman, A., Bennett, V., Friend, C., et al., 2016. Nature 537, 535.

Peacock, S., Barman, T., Shkolnik, E.L., et al., 2019. ApJ 871, 235.

Peacock, S., Barman, T., Shkolnik, E.L., et al., 2020. ApJ 895, 5.

Pearce, B.K.D., Pudritz, R.E., Semenov, D.A., et al., 2017. Proc. Natl. Acad. Sci. U. S. A. 114, 11327.

Peterson, B.M., 1993. PASP 105, 247.

Powner, M.W., Gerland, B., Sutherland, J.D., 2009. Nature 459, 239.

Ribas, I., Gregg, M.D., Boyajian, T.S., et al., 2017. A&A 603, 58.

Sachkov, M., Sichevsky, S., Shustov, B., et al., 2020. Proc. SPIE Space Teles. Inst. 2020: UV to Gamma, 1144474.

Sanz-Forcada, J., Micela, G., Ribas, I., et al., 2011. A&A 532, A6.

Schneider, A.C., Shkolnik, E.L., 2018. AJ 155, 122.

Seaton, 1979. MNRAS 187, 73.

Segura, A., Walkowicz, L.M., Meadows, V., et al., 2010. Astrobiology 10, 751.

Shkolnik, E.L., Barman, T.S., 2014. AJ 148, 64.

Shustov, B., Sachkov, M., Gómez de Castro, A.I., et al., 2018. Ap&SS 363, 64.

Sing, D.K., Pont, F., Aigrain, S., et al., 2011. MNRAS 416, 1443.

Sing, D., Lavvas, P., Ballester, G., et al., 2019. AJ 158, 91.

Tilley, M.A., Segura, A., Meadows, V., et al., 2019. Astrobiology 19, 64.

Tu, L., Johnstone, C., Gudel, M., 2015. A&A 577, 3.

Vidal-Madjar, A., Lecavelier des Etangs, A., Desert, J.-M., et al., 2003. Nature 422, 143.

Vidal-Madjar, A., D esert, J.M., Lecavelier des Etangs, A., et al., 2004. ApJL 604, L69.

Wong, M.H., Simon, A.A., Tollefson, J.W., et al., 2020. ApJS 247, 58.

Index

'Note: Page numbers followed by "f" indicate figures and "t" indicate tables.'

Printed in the United States
by Baker & Taylor Publisher Services